THE WHISTLING HUNTERS

In the SUNY Series in Animal Behavior
Jerram L. Brown, Editor

MICHAEL FOX formerly a professor of psychology at Washington University, St. Louis, Missouri, is currently Director, Institute for the Study of Animal Problems, The Humane Society of the United States, Washington, D. C. He holds B. Vet. Med., from the Royal Veterinary College, London, and PH.D., and D.Sc. degrees from the University of London and is a member of the Royal College of Veterinary Surgeons, England. He is contributing editor to *McCall's* magazine and writes a syndicated newspaper column *Ask Your Animal Doctor*. He has written several popular books, including *Understanding Your Dog, Understanding Your Cat,* and *Between Animal and Man: The Key to the Kingdom*, and also technical books such as *The Behavior of Wolves, Dogs and Related Canids, The Dog: Its Domestication and Behavior, Concepts in Ethology: Animal and Human Behavior*, and *Integrative Development of Brain and Behavior in the Dog*. He has also published several children's books, notably *The Wolf, Sundance Coyote,* and most recently *Ramu and Chennai*, a story about these whistling dogs and a jungle boy.

THE WHISTLING HUNTERS

FIELD STUDIES OF THE ASIATIC WILD DOG (*Cuon alpinus*)

Michael W. Fox

STATE UNIVERSITY OF NEW YORK PRESS
Albany

Published by
State University of New York Press, Albany

Printed in the United States of America

For information, address State University of New York Press, State University Plaza, Albany, N.Y., 12246

Library of Congress Cataloging in Publication Data

Fox, Michael W., 1937-
The whistling hunters.
(SUNY series in animal behavior)
Bibliography: p.
Includes index.
1. Cuon alpinus I. Title II. Series.
QL737.C22F6935 1984 599.74′442 84-2442
ISBN 0-87395-842-X
ISBN 0-87395-843-8 (pbk.)

CONTENTS

FOREWORD

The jungle whistling dog, or dhole, a distinct and separate species of the canid family, is Asia's virtually unstudied wild dog. Today, it is one of many endangered species threatened with extinction. Highly intelligent, wary of man, and an elusive predator in the dense jungle of Southwest India, this rare and beautiful creature is one of the most difficult animals to study. Its very nature defies the patience and skill of the most dedicated naturalist; yet, knowledge about its habits and a more widespread understanding and appreciation of this species are essential for its protection and continued survival.

In this book I bring together fragments of my own field studies along with the few observations other naturalists have made. Thus I assemble as complete a picture as possible of the whistling dog. The exquisitely complex mosaic of its life and place in the wild may well be shattered and lost forever, because wild dogs may soon become extinct if their role in nature is not respected and their habitats protected. Extinction is forever, like a stargate closing for all eternity. It would mean the end, not only of wild dogs, but also of the jungle wilderness itself, for wild dog and its habitat form an inseparable whole, a distinct and unique presence. The jungle is their spirit of place. It is this spirit what René Dubos* calls "a God within", that I

*Dubos R., *A God Within* (New York: Scribner, 1983).

have attempted to comprehend, as well as the wild dogs' natural history. I have also sought to understand wildness—the essence of wilderness—in this study, since it, like the whistling dog, is no less threatened with extinction.

This book is, therefore, something of a mosaic in itself, attempting to convey more than an understanding of the behavior and ecology of the dhole (a subject which requires considerably more field research), but also of how field work is conducted. The trials, tribulations, and more subjective impressions that field researchers experience in the few and precious wilderness areas left on earth today are as important to share, I believe, as the objective data and theoretical conclusions of such field research. The urgency in protecting biologically unique and spiritually inspiring places is too often understated. Politics, agricultural and industrial activities, philosophy, and even religious attitudes are germane to the study of wild dogs (as they are to any wild creature), since these factors indirectly and, in some instances, directly impinge upon their behavior, ecology, and future survival.

Purely objective (scientific) knowledge, although of value, is in itself limited, for, as the Indian *Bhagavad Gita* aptly states, "When one sees Eternity in things that pass away and Infinity in finite things, then one has pure knowledge. But if one merely sees the diversity of things, with their divisions and limitations, then one has impure knowledge." Thus, while recognizing the limited, if not "impure," knowledge available on the ecology and behavior of the Indian wild dog, this book endeavors to convey something of the sense of Eternity and Infinity that is an integral, subjective aspect of the natural history of *Cuon alpinus* in its jungle habitat.

> "The immortal is veiled by the real. The Spirit of life is the immortal. Name and form are real, and by them the Spirit is veiled"
>
> —from the Brihad-Aranyaka Upanishad.

Michael W. Fox
Washington, D.C.
June 1983

INTRODUCTION

The Asiatic wild dog *Cuon alpinus* is a little-known, distant cousin of our domestic dogs. Like the wolf and Cape hunting dog of Africa, it hunts in packs and, like so much of the world's wildlife today, is an ever-diminishing, endangered species. My earlier research with dogs and with captive wolves, coyotes, jackals, and foxes led me to study this missing link in the field in order to complete a conceptual picture of the social behavior, ecology, and evolution of canids. Although the wild dog was the "target animal," studying it was impossible without also studying all the complex interrelationships which impinged upon it either directly or indirectly in its jungle habitat. The study area chosen was ideal—the southwest of peninsular India, where correspondent and local naturalist E. R. C. Davidar had studied these animals over several years.

I undertook an intensive, preliminary, two-month study, assisted by an experienced Indian field zoologist, A. J. T. Johnsingh, between January and March 1974. The time was ideal. It was the spring dry season, when the wild dogs would be rearing cubs and vegetation would be sparse so that the animals might be observed within minimal cover. I subsequently conducted a second field study between October and December 1975, and field studies were continued by Johnsingh in Bandipur Sanctuary in the same district for a further two years. Some of the data from Johnsingh's doctoral thesis will be referred to in this book.

The jungle gave me something more than simply new scientific data on *Cuon alpinus*; it provided a view of nature, of wilderness and wildness in which the mind and spirit of man could be an integral and natural part. My search for wildness and whistling dogs had two purposes. First, my quest was to learn more of these pack hunters and of the world in which they live, which is today endangered and may not exist in fifteen to twenty years. Second, my trip was a personal, inner search for meaning, and the jungle was both my teacher (or guru to be geographically correct) and my place for reflection (or meditation if you wish!). I have combined both objective facts and subjective views in the ensuing chapters to provide a first-hand account of the jungle, the wild dog's habitat, and I have included most of what little has been written about this Pleistocene hunter which once abounded in Europe as well as Asia. This material necessitates critical evaluation from the standpoint of conservation and humane ethics. It is to these objectives, and to the whistling dog, that this work is dedicated.

Chapter 1

THE FIRST DAYS

The study area was located in a wide valley nestled in the Nilgiri Hills, an extension of the Western Ghats mountains in southwestern India. At an elevation of 3,000 feet, this valley contained several small villages, an expanding agriculture, and some virtually pristine areas of jungle habitat (designated the Avaraihalla Reserve Forest) some 15 kilometers north of the town of Ootacamund (Ooky) on the Madras side of the Madras-Mysore border formed by the Moyar river. The area studied was a jungle sector of approximately 25 square miles near the villages of Masinigudi and Mavanahalla. It included protected land within a 250-square-mile sanctuary (Mudumalai) and an adjoining hunting block abundant with wildlife managed by the Nilgiri Wildlife Association.

The entire valley was an undulating expanse of dense thorn bush, lantana, and laburnum shrubs, as well as deciduous trees such as wild fig, mango, and tamarind. There were countless small streams or nullahs and two larger rivers, the Sigur and Moyar. The streams became parched creeks during the dry season—between December and March. Giant bamboo lined the banks of the larger water courses; in many areas, the dense scrub gave way to more open, parklike expanses which were usually heavily grazed by domestic cattle. The entire valley was a macrame of narrow, meandering trails, and, but for the distant hills, one unfamiliar with the area

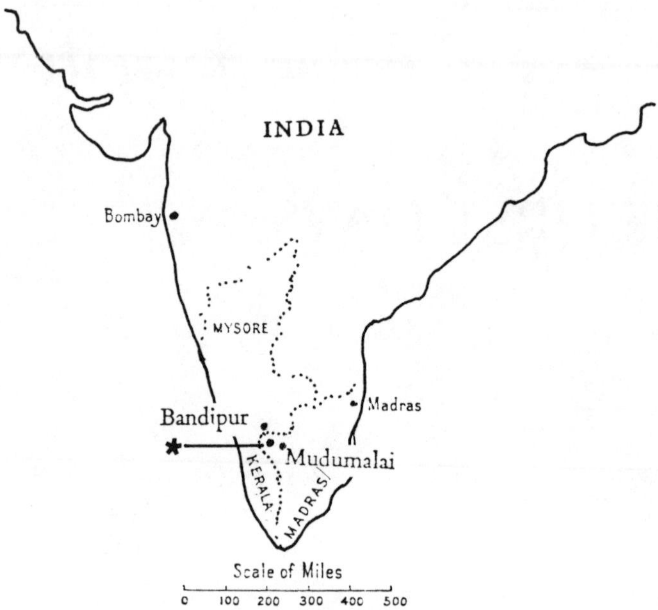

Figure 1.1 Study area shown on map of India

would have little sense of direction and could easily get lost. The dense vegetation and undulating terrain usually limited visibility to only a few yards, which made direct observation of wildlife extremely difficult and wildlife photography a frustrating and often hopeless endeavor, unlike the open plains of Kenya and Tanzania.

I had come a long way to study one of the most elusive of hunters. Trappers in America had told me that even after twenty years in the wilderness they had rarely seen a wolf, and I was after its distant cousin, the whistling dog, in a place where even an elephant was often difficult to see. The jungle would give the dogs so much cover I would be lucky even to see them, and a fleeting glimpse would hardly tell me much about their behavior. Unlike the Cape hunting dog of Africa, which is easy to approach, photograph, and study, the whistling dogs would flee like the wolf at the sight or scent of a man.

Since reading Kipling's story of the Mowgli and these wild dogs he called the "red peril" of the jungle, these animals had captured my imagination. My research on the wolf and on other canids of North America, such as the coyote and the red, grey, and Arctic foxes, left many unanswered questions. A field study of one of their relatives, the wild dog, would help.

Only the wolf, the whistling dog, and the African or Cape hunting

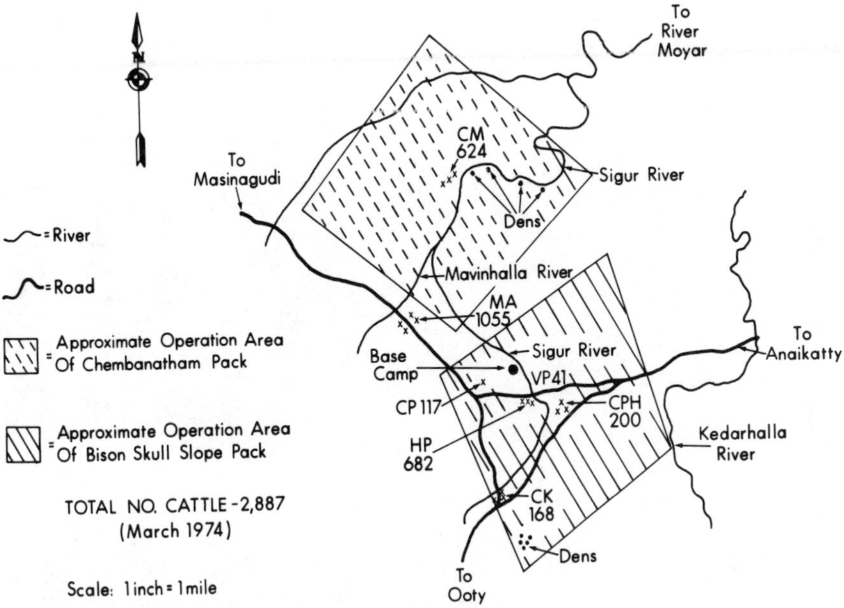

Figure 1.2 Study area in Nilgiri Hills, adjacent to Mudumalai and Bandipur wildlife sanctuaries. (Initials and numbers refer to cattle farms and number of stock, respectively.)

dog hunt in organized packs. Anthropologists and ethologists alike have emphasized that studies of such hunters may add more to our knowledge about the biological roots of human behavior than field work on primates such as baboons and chimpanzees. The latter are primarily vegetarians and rarely hunt. But man was once a hunter and has closer affinities with the wolf than one might imagine. To hunt effectively in a group, animals must possess an efficient means of communication and a high degree of intelligence to coordinate pack activities. With weapons—teeth or spears—there must be some rules to prevent killing and to reduce the chances of conflict within the pack or tribe. Hunting must not be excessive, nor the growth of the hunting population too great; otherwise, overkill and overpopulation will have catastrophic ecological consequences.

We have already learned much from the wolf (Mech 1970) and the African hunting dog (Van Lawick and Van Lawick Goodall 1970 and James Malcolm, personal communication 1981) as well as some facts and provocative theories about man the hunter (Lee and DeVore 1968), but virtually nothing is known about the other pack hunter, the whistling dog (Cohen 1977). How do these animals communicate in a dense jungle and coordinate their hunts? Do they really hunt in packs of thirty or forty and sometimes kill leopards and tigers? Do they ever attack humans and are they the bloody killers of legend?

Figure 1.3 Parklike areas reflect overgrazing by domestic livestock. Typical terrain is dense scrub with small clearings and countless interconnecting game trails (bottom photo), making observations of wildlife extremely difficult.

Intellectual curiosity was not my sole motive for going so far to study, what may seem on the surface, so little. I have found that if I can get into another animal's "space"—its world, its ecosystem, and its behavior—I can learn something about myself as well as about life itself. The essence of my popular books *Understanding Your Dog* and *The Soul of the Wolf,* and of my children's books such as *The Wolf, Sundance Coyote,* and *Ramu and Chennai,* comes from a synthesis of scientific facts and feeling or empathy, if you wish. The combination of these two elements (of objectivity and subjectivity) in the study of animals' behavior can, I believe, bring us closer to understanding some facets of our own nature and behavior, as well as engendering a deeper reverence for animals.

These ideas were on my mind as we walked along a narrow trail between an endless thicket of thorn bushes, wild fig, mango, and tamarind trees in the valley between the blue Nilgiri hills. The sun had not yet risen above the distant hills, and the cold crept through my khaki jacket that would not survive the intermittent attack of thorn bushes. At least there were no flies: it was the dry season and the ground was baked hard by the sun. The earth on the trail was a fine dust, pounded by the feet of wild animals, especially herds of small chital (or spotted deer) and large sambar (or giant deer). We were following the tracks of two wild dogs clearly visible on the trail, made earlier that morning. The tracks would "age" as gentle breezes smoothed the powdery dust that now held a perfect impression of the dogs' feet. Perhaps these two were out on their morning hunt.

Johnsingh, a zoology lecturer on leave from Sivakasi College, was my field assistant, and ahead of us was our tracker (or shikari), Sikka Bunda. Bunda could tell a story from a swirl of dust on the trail and his keen eyes could spot an animal under cover that I could, at first, see only with binoculars. I would have to grow up in this region and spend forty years in the jungle before my senses could ever match his, I thought.

Bunda stopped, and with a slight movement of his machete pointed toward a dense clump of bushes. Before I could raise my camera, a giant sambar stag broke cover, ran across the trail in front of us, and vanished.

We walked with the wind in our faces so that any animals ahead would not smell us and flee before we had a chance to see them. More importantly, we would have a better chance of avoiding lone elephants—solitary males who were extremely unpredictable and quite dangerous. By 10:00 A.M. the valley was extremely hot, and we chose a place by the river for breakfast. That morning we saw no wild dogs, but I learned to read the tracks and signs of wild pig, mongoose, sloth bear, sambar, chital, peafowl, elephant, jackal, and porcupine. The valley was, indeed, abundant in wildlife, but I did wonder if all these animals really existed, since all I had seen was one sambar.

The next day my senses seemed more in tune and each day I saw and learned more about the inhabitants of this incredible land. Early morning was filled with sounds that eventually had identity: the piercing screams of lapwings and catlike mews of peacocks parading with emerald tails three or four feet long. In the bamboo that rose twenty or more feet above the river, common grey langur monkeys hooted and bonnet macaques chattered and

Figure 1.4 Except for coat color and larger size, the African or Cape hunting dog (upper and lower right) closely resembles the dhole or Asiatic whistling dog (upper and lower left). (Upper right photo courtesy of James Malcolm)

tumbled. Sounds like the cracking explosions of firecrackers identified an elephant tearing the bamboo with its powerful trunk. A trumpetlike scream or a resonant whistle meant a sambar or chital had been alarmed, possibly by wild dogs. The whoop of a langur, the machine-gun chatter of a Malabar

squirrel in the bamboo, or the cackle-scream of jungle fowl were also alarm signals. Deer drinking by the river would immediately flee for cover if they heard any one of these sounds. Sometimes it was a false alarm—one squirrel would simply be threatening a rival in an adjacent stand of bamboo—but still the deer fled!

I learned to listen to these sounds, as well as to the silence, since they could help us locate the wild dogs. We also "read the skies" for signs—a

Figure 1.5 Tall clumps of bamboo stand at the edge of rivers meandering through the wide valley, which is covered mainly by deciduous forest.

flying, cawing jungle crow or circling vultures might lead us to the dogs or to their kill.

We adopted a work pattern of systematic tracking in the mornings, from sunrise until 10:00 A.M., and sometimes again in the late afternoon. We would spend the rest of the day in ground or tree hides strategically located either near a trail frequented by the wild dogs, near a denning area, or at a favored watering place of deer. Sometimes we would sit all day and wait. Long hours alone in a hide do strange things to the mind and to the body: you discover the human frame is not built for sitting eight or more hours without moving!

Patience and determination are usually rewarded, but I began to wonder when. After six days we had seen no wild dogs. Johnsingh, who had already spent a full month in the area, had seen them only four times. The day prior to my arrival he had been treed by a wild dog! His own fear of the dog probably put him up the tamarind tree, since there is no record of a wild dog attacking a man. He had come across a dead cub, in what we later found to be a temporary nursery, and its mother suddenly appeared. She growled and yipped at him and he scrambled up the tree as the bitch ran toward where he had been standing, toward the dead cub.

On my first day we found some wild dog tracks. On the second day we came across a communal defecation area where forty or more wild dog droppings had been deposited at an intersection of two trails. This was a curious phenomenon, and we certainly hoped to learn more. Perhaps this behavior could be some kind of territorial marking, like the "pasting" ceremonies Hans Kruuk (1972) describes in the African spotted hyena. The third day we visited an abandoned wild dog den, and on the fourth day we found a fresh kill. The dogs had killed a chital stag no more than 200 yards from our camp, while we were tracking them on the other side of the valley. That night we ate a fine venison curry, and I felt close to the wild dogs. I had tracked them, collected their droppings, visited their den, and now even shared their food.

Before dinner the next evening I went to one of the tree hides by the river. As I settled into the hide, or *machan,* I noticed my intrusion was like a stone falling into water. A ring of silence grew around me; the birds, frogs, monkeys, and other animals were silent. After a while, a fan-tailed flycatcher began to resume its cheeping and soon other birds were singing and frogs were croaking again; the hole my presence made had healed. The sounds were almost mesmerizing. I could listen to just one frog croaking, then another, and then let my consciousness slide and be swallowed up by the undulating cacophony of myriad frogs. Each croaked from atop its stone and, in the growing darkness, the stones seemed to come to life, each with its own distinct voice but uniting with others in the same incessant rhythm.

My mind was wandering again, half asleep, or gone to that netherland it visited whenever I settled in to a *machan.* Suddenly, the trumpeting alarm of a sambar blasted close by and soon a half-whistle, half-toot, came from the far bank, repeated a dozen times. It was the incredible but distinctive call of the whistling dog. Surely, I thought, the following day I will see one.

We spent most of the next day in a ground hide overlooking a whole complex of wild dog dens in a rocky outcrop some 100 feet above the plains. The only life we had seen there was a pair of mongoose early in the morning and an elephant making its way through dense scrub, a visitor we were lucky to avoid. I was watching an army of ants busily conducting its business in a never-ending trail up and down the tree while I wove the fourth bracelet of vines for my daughter. I had at last figured out how the Africans make those beautiful elephant-tail hairs into bracelets with two knots and no ends showing. A children's story I was planning to write, about a wild dog and an Indian boy, was also occupying my mind. It is surprising how the hands and mind become so active and creative when you just sit: creative boredom.

Suddenly, and for no particular reason, I decided to whistle. I blew four times, like the wild dog I had heard: a call for assembly. I whistled again, remembering my friend Reggie Davidar's statement that you can

Figure 1.6 In the dry season, even the largest rivers become rocky ribbons of shrinking pools, and at such watering holes observation hides are invaluable.

whistle on a 35-caliber rifle cartridge and call the wild dogs. Bounty hunters used to shoot wild dogs easily this way.

My peripheral vision had included the denning area all day and now I was beginning to see things, projections of a mind expecting something. Then, one appeared, a short-legged animal, sniffing the ground we had walked earlier that morning. It had responded to my call! Its rich, almost mahogany-colored coat glistened in the sun, the rays of light splitting on the short dark hairs of its back to create a halo of silvery fibers. Then a cub appeared from the bushes, hesitantly entering the clearing behind the larger dog. Before my camera could record a fraction of what I saw and felt, the animals had disappeared.

At sundown the two mongoose came out again, and a coppersmith began its repetitive song from a nearby tree. We went back to the camp elated, but knowing that only with systematic and continuous effort, as well as an element of luck, could we hope to learn anything about these elusive wild dogs.

Studying the whistling dogs, even in the dry season when conditions were optimal, was extremely difficult. Alert, wary of human presence and often active even before sunrise, they made direct observation difficult and chancy. Observations around den sites were virtually impossible because the dens were usually located in rocky outcrops or deep, dry creek banks covered with dense shrubbery. Our close proximity might be detected by an alert dog, and even if the wind was in our favor a bird or monkey could see us from the trees and alert half the valley with an alarm call.

Still hoping to learn more of the dogs' social lives, a second field study was completed in the same area between October and December 1975. This period, after the rains, was chosen to coincide with the wild dogs' breeding season. They would be less mobile if encumbered with one or more litters and therefore confined for some weeks to a denning area. With an additional three field workers,* we would be able to glean much more than the few pieces of information the initial study yielded.

After the rains, the jungle was a very different place. Tall grass, leafy bushes, and full rivers and water holes resulted in widely dispersed wildlife. Visibility and direct observation were even more limited. Prey and predators alike were no longer restricted to the greenery and water pools along the drying rivers, and now had the vast undulating valley and all its water at their disposal.

In October, the northeastern monsoons returned, renewing the vegetation already luxuriant after the long rains earlier in the year. The monsoons not only limited visibility, but they rendered many of the areas we had visited during the dry season unrecognizable and, at times, inaccessible. River beds that were once rocky roads with occasional pools became impassable, raging torrents. One advantage of the rains, though, was that the muddy trails of animal tracks were cleared each day. With all recent history erased, tracking was greatly facilitated since only fresh signs after a brief downpour were evident.

And now a new element—teeming life—was introduced. In the dry season, many living things laid dormant in seed, egg, cocoon, or chrysalis. Now, every individuated life form was active, streaming, crawling, blossoming, and buzzing. A river of life flowing from the late Pleistocene period and before into the teeming present: the shapes, smells, colors, and sounds unchanged in their own replicated forms yet each unique in its individual expression of ancestry. Individual, species, genus, family, phylum—of the past and of the present . . . and with a future uncertain.

This small oasis in the blue hills was a living museum of beauty and integrated diversity. Each living thing was a testament to the incredible phenomenon of evolution, each was a successful experiment in survival and procreation. Each individual was subtly attuned to others, interdependent, interpenetrating—a life mosaic of myriad forms, the essence of which perhaps only humans are capable of appreciating. Humans can enjoy and enjoin nature spiritually as well as physically, and only human beings can preserve or destroy all that still is.

---

*My wife Deborah and two students from Washington University, St. Louis, Bruce Barnett and Jim Cohen.

Figure 1.7 Lush vegetation after the rains contrasts with the defoliated and overgrazed scrub jungle in the dry season.

Gone are the magnificent *megafauna* of the Pleistocene, but here elephant, hyena, deer and tiger, wild dog and sambar survive: evolutionary successes. Past ages of giant insects, plants, and reptiles are now echoes in miniature—dragon flies, social ants, bees, termites, small lizards, toads, chameleons, and butterflies of every hue, winging iridescent signals between orange, pink, and red bushes of lantana and yellow Indian laburnum.

Chapter 2

JUNGLE SENSES

One's senses slowly become attuned to the jungle as the rhythms of mind and body are reset to the time of the chennai, the whistling dogs, and one becomes part of their spirit of place. Shapes and shadows created by light and moving trees, once illusory forms, cease to be given life by the imagination. The mind and eye begin to see through this protective canopy and even through the subtle camouflage of chital, sambar, and chennai. Finally, the proximity of life can be felt and the senses focused with accuracy.

Motion is no less illusory. At first, a glimpse of some distant object (a red termite mound, for example) between the trees and shrubs while one is moving gives the impression it is also in motion. Eventually, the eye and brain correct for the motion of the body and no longer see movement in inanimate objects. Then, the movement of animals through the dense jungle can be quickly detected and separated from the background.

The silence of one's presence is a warning to all creatures: when birds stop singing the message is of intrusion and potential danger. The problem is like trying to walk on snow while leaving no imprint—no matter how carefully you move, the mark is made. The silence is as though a wound has been inflicted on the jungle by human presence. So, the observer must be still and let the jungle come to life around him again. Stillness and patience are soon rewarded when one, then more birds begin to sing and the hole of silence quickly heals.

At other times, human presence is widely broadcast by the alarm calls of birds as well as by deer and monkeys. But these animals also occasionally give alarm calls when a whistling dog or panther is close, and so they indirectly may help the observer. Unfortunately, this service is rare and one is usually forced to remain concealed in a hide.

A hide, however, is a poor substitute for invisibility. Even though one cannot be seen or heard (until restriction of circulation or nerve pressure necessitates some movement), there is no way to prevent animals from picking up one's scent. Masking it with deer musk or elephant dung appealed more to the scientific mind than to the senses and was dismissed on aesthetic grounds and with a prayer to the wind. One can not hope to be invisible when surrounded by a jungle of eyes, ears, and noses better attuned than one's own. At best, one can only try to keep a "low profile" and hope that, with repeated exposure, some animals will become habituated to a curious biped in jungle green.

Information travels quickly in the jungle, information not only of one's presence or the presence of a predator, but also of a whistling dog hunt. Alarm calls of would-be prey or the scream of prey when attacked alert the jungle crows, whose excited calls bring vultures from miles around. The latters' presence, circling the skies, is another piece of information telegraphed through the jungle, and other scavengers—jackal, hyena, and village pariah or pi-dog—close in. Finally, the smell of blood and torn flesh attracts carrion flies, which lay their eggs in the prey's remains after other scavengers have left.

The jungle is, therefore, not only a complex of many diverse and interdependent plants and animals, but also an intricate communication network to which the human mind can gradually become attuned. Different sounds identify different species, and express emotion and intention often serving to repel rivals, to warn companions, or to attract mates. The totality of sounds merges separate identities into one, an oceanic rhythm that penetrates the entire jungle. Your ears become attuned to this rhythm, identifying individual sounds easily and then returning to attend that integrated background of sound. Any change in its pattern—a hole of silence somewhere, or a distant alarm call—can be detected immediately. To concentrate on the individual sounds is to lose the totality of it all. Conversely, to be lost in this ocean of sound is to miss hearing the changes that can convey so much information.

The sense of touch also becomes attuned to the jungle. Your body can sense the direction of the wind, while the sensation of warm sunlight on one side of you provides a constant navigational reference. Rocks and twigs that roll or snap beneath your feet are carefully felt before the earth receives the full weight of your stride. I longed to be able to move with the tender silence of bare feet; stealth in boots is difficult to acquire. Even elephants move with impressive silence, but not I. The noises that my unfeeling boots

Figure 2.1 Two of our allies—the jungle crow and the vulture—whose behavior assisted us in locating wild dogs and their kills.

so often created was as alarming to the animals as it was to my own senses. Yet, I had no alternative but to wear those boots and to walk as though I were treading on bamboo thorns with bare feet, as silently as I could.

On those rare days when there was a breeze, my noisy feet had an ally: creaking bamboo, rustling leaves, and groaning branches muffle the sounds of leaves and twigs cracking underfoot. With the breeze in your face, and so in your favor, such days make you feel almost invisible, although animals seem to be even more wary at such times. It is quite remarkable how the breeze carries sounds in its prevailing direction: a river, no more than a background noise in the distance, suddenly sounds closer. As the dry season progresses and shrubs and trees lose their leaves, sounds travel more easily. In some areas of the jungle, noises have a resonant, bell-like quality which, during the rains, do not carry far because of the thick vegetation and so are mellowed.

After four or five weeks in the jungle, having perhaps seen a whistling dog for no more than a few seconds in spite of hundreds of hours searching and waiting, my mind started doing strange things to my senses. As in sensory-deprivation experiments where people are denied any input, my mind started to work like a movie projector. Instead of those earlier illusions of shape and shadow, I began to suffer real delusions. Expecting and hoping to see wild dogs, wild dogs seemed to appear—out of the shapes and shadows of rocks and leaves, in open glades, and on top of rocky slabs. Long hours of sitting in a *machan* may have produced this intriguing sensory delusion; certainly, if the wild dogs had been more numerous, we would not have suffered from this deprivation syndrome! A more usual mind game was to lapse into a very vivid illusion (such as talking to someone in a living room or office somewhere). This dreamlike state went on while eyes and ears remained fully alert and attuned for the possible appearance of animals near the hide.

The delusional state (wherein wild dogs seemed to appear) is, I believe, a major underlying cause for many hunting accidents, especially at the beginning of the season in the United States. A nervous system, controlled by a mind with a "search image" of a deer, responds to any movement, color, or shape detected by the mistuned senses. A hunter in a red jacket is shot (because movement meant a deer to the deluded mind), or a child in a fawn-colored sweater is shot off her bicycle (because the mind projected the image "deer" onto the color). Little wonder there are so many hunting accidents.

Although the eye can quickly learn to track, following footprints that (by the clarity of their imprint) are obviously fresh, a rudimentary sense of smell is a constant frustration. Knowing how sensitive other animals are to wind-borne odors, I adopted the strategy of the wild dog: traveling with the wind in my favor. Then I had a more equal advantage, since the animals ahead would only be able to see or hear the human intruder. With time, I could see almost as well as they.

Wild dogs and other carnivores also travel crosswind. By moving across rather than into the wind they are able to scan a long transect for prey. If the prey's scent is picked up, the predator orients directly into the wind and tracks it down. With such poor noses, however, this is no strategy for humans.

Although hunting does not always lead to success for the wild dogs, adopting some of their strategies when observing prey is essential. But then,

Figure 2.2 With mottled light and shade, the jungle not only makes photography a challenge but becomes a dangerous place for the unwary. You might easily miss an elephant (above) resting in the shade on a hot afternoon—or if scared, you could imagine elephants everywhere!

what is the best strategy for hunting the hunter? The chennai are infinitely superior hunters and it is hard to come close to them either in ability or proximity.

Part of the frustrating elusiveness of the dogs lies in their hunting pattern. The chennai increase the probability of a successful kill by randomizing their daily activity, first hunting in one area, then another, in a pattern wholly unpredictable to the observer and, most probably, to the prey as well. So I decided to adopt this pattern (i.e. endeavor to think and act like a wild dog) and, by sheer chance (or luck), I might happen on a kill. Persistently visiting one area, consistently visiting different areas at regular intervals, or just sitting in a hide by a well-used track or water hole are all viable alternatives. All strategies boil down to luck in the end, since the chennai are so predictably unpredictable—the mark of a superior hunter.

Walks in the jungle gradually bring mind and body together so that one moves and senses things in perfect synchrony, outer and inner realities becoming the same. There is no time or reason to be preoccupied in abstract thought. Such could be hazardous, because of the ever-present danger of elephants. The light breeze and sunshine, or incessant rain, maintain continuous contact. Movement and sensation without distracting thoughts ebb and flow like a constant meditation, and one becomes transparent, open to the jungle and part of the numinous spirit of place.

One evening, I found a rock, a tamarind seed, and a feather in my pockets. Why had I picked up these things, of no obvious value or scientific significance? These things held meaning for the senses, opened by the jungle while mind and body were one with everything around it. Was there something still alive and tangible in these natural objects which resonated with the senses and to which the mind, in turn, gave meaning? Rock, primordial essence incorporated into all living structures, returns to the soil and reforms: an endless cycle. Seed embodies potential, fruit and embryo, the end and the beginning, a link in the continuum of individuated life. Feather, product of eons of evolution, designed from a genetic blueprint, exquisitely expresses fulfillment through adaptation; structure, function, and environment are unified. Organism and environment are one and inseparable.

Each treasure, with its own truth, each a lesson from nature, opens the mind to the totality of it all. Each object, unique in its form, is like a point in time, a droplet from the endless river of life. The cadence of life in the jungle and the rhythms of separate lives follow patterns similar to the flow of streams and countless tributaries. From the infinite springs of undifferentiated potential to the protean ocean of consciousness, the river of life flows, breaking into myriad streams and tributaries. In these separate paths, the rate of flow, through evolutionary time, varies for each phylum, genus, family, and species. Each stream has its own rhythm or rate of flow. Some cease to flow and their extinction affects the flow and branching of other streams and tributaries, and the total pattern gradually changes. Also, mutation, the cadence of flow within each life form, continues; identity is preserved through the generations. Balance and harmony are maintained, since all separate tributaries are interconnected—what the American Indians refer to as the sacred web of life.

These things you come to sense intuitively as you become attuned to the diversity of life that abounds in the jungle. This transformation of the human mind, bizarre or sophomorically poetic as it may sound, is perhaps the first essential step before you can enter and begin to understand the whistling dogs' world.

Chapter 3

THE WHISTLING DOG'S WORLD

The village of Masinigudi, four miles from our camp, served as our supply and information center. In any village, everyone knows everybody's business, and everyone, from farmer to forester, from baker to bamboo cutter, knew why we had come. Every other day we had at least one "informant" with some news about wild dog activity. These tips occasionally helped us locate kills and, especially, enabled us to estimate the size of hunting packs by the number of dogs seen together. Interestingly, the informants were usually accurate in counts up to ten, then they would estimate from "twenty to forty dogs" and "twenty-five to a hundred." Perhaps a group of fifteen or twenty dogs milling around, in and out of the bushes, is extremely confusing and you get the impression there are many more dogs than there really are. Thus, the reliability index for numerical counts drops dramatically with increasing numbers of dogs in a pack or assembled clan. One man claimed he was once surrounded by a thousand dogs—obviously human perception is effected by the confusion created by an actively milling clan of wild dogs. This confusion undoubtedly contributes to the myths surrounding this social hunter!

The study area was replete with a wealth of varied species of bird. The following is a short list of those frequently observed: yellow-headed and pied wagtails, fantailed and paradise fly catchers, green bee eater, hoopoe, shama, black and racket-tailed drongos, common babbler, blossomheaded parakeet, roseringed and large green parakeets, crow-pheasant, grouse, partridge, peafowl, jungle fowl, grey and Malabar hornbills, sunbirds, kingfishers, golden-backed woodpecker, golden oriole, curlew or lapwing, bulbul, Brahmini myna, crested serpent eagle, crested hawk eagle, pariah kite, coppersmith, common green pigeon, scarlet minivet, Indian blue jaw, Malabar whistling thrush, Indian robin, jungle crow, common vulture, king vulture, white scavenger vulture, shrikes, common sparrow, nightjar, horned owl, and sparrow hawk.

These birds were a delight to see and hear: the eerie cries of peafowl and lapwing early in the morning, and the incessant racket of jungle babblers, always in noisy groups of six or seven. Parakeets, shooting like colored rockets through the trees, would shreak and chatter constantly. The humanlike, morning melodies of the whistling thrush and hot afternoon rhythms of the coppersmith would later give way to the evening chorus of countless birds, a sound as in tune with that time of day as the more exuberant babble, twitter, and trill of the dawn chorus.

The study area was also rich in diverse species of mammals,* knowledge of which was essential for an understanding of the ecological interrelationships between prey and predator. A general index of their relative abundance is shown in Table 3.1.

Table 3.1 *Mammalian Species Identified in Study Area*

Species	*Abundance*
Cervidae	
Chital (*Axis axis*)	+ + +
Sambar (*Cervus unicolor*)	+ +
Barking deer (*Muntiacus muntjak*)	+
Tragulidae	
Mouse deer (*Tragulus meminna*)	+ + +
Bovidae	
Gaur (bison) (*Bos gaurus*)	+
Elephantidae	
Elephant (*Elephas maximus*)	+ + +
Suidae	
Wild pig (*Sus scrofa*)	+ + +
Leporidae	
Indian hare (*Lepus nigricollis*)	+ + +

*Recent population surveys of the adjacent Mudumalai sanctuary reveal the following approximate figures within its 123 square miles of deciduous jungle: 350 elephants (including 60 tuskers), 50 panthers, 12 tigers, 1,750 chital, and 12 herds of sambar (2–4 per herd). The elephant population is obviously high and control measures to regulate this species, as was recently instigated in some African sanctuaries, will be needed.

Table 3.1 *Mammalian Species Identified in Study Area*

Species	*Abundance*
Hystricidae	
Indian porcupine (*Hystrix indica*)	+ + +
Scivridae	
Malabar squirrel (*Ratufa indica*)	+ + +
Small flying squirrel (*Petinomys fuscocapillus*)	+
Palm squirrel (*Funambulus palmarum*)	+ + +
Primates	
Common gray langur (*Presbytis entellus*)	+ + +
Bonnet macaque monkey (*Macaca radiata*)	+ + +
Cricetidae	
Field or Bush rat (*Millardia meltada*)	+ + +
Herpestinae	
Common mongoose (*Herpestes edwardsi*)	+ + +
Viverridae	
Palm civet (*Paradoxurus spp*)	+
Manidae	
Indian pangolis (*Manis crassicandata*)	+
Ursidae	
Sloth bear (*Melursus ursinus*)	+
Felidae (cats)	
Leopard (*Panthera pardus*)	+
Jungle cat (*Felis chaus*)	+
Tiger (*Panthera tigris*)	+
Mustelidae	
Common otter (*Lutra lutra*)	+ +
Clawless otter (*Anoyx cinerea*)	+ +
Canidae	
Jackal (*Canis aureus*)	+ +
Wild dog (*Cuon alpinus*)	+ + +
Hyaenidae	
Striped hyena (*Hyaena hyaena*)	+ +

Estimates based on direct daily observations and indirectly from tracks, scats, and vocalizations: + + + common
+ + occasional
+ rare

Always active somewhere in the canopy of trees and tall bamboo were bonnet macaque monkeys *(Macaca radiata)* and common langurs *(Presbytis entellus)*. These fascinating primates were invariably seen in large family groups or troops. Both species were extremely wary of humans and when an observer was detected by one or more alert individuals, the entire troop would rapidly disappear through the treetops, crashing noisily and uttering alarm calls as they went. They were even more wary when on the ground, foraging for insects, small lizards, and fruit from lantana and zizyphus. The langurs carried their long tails high in an inverted "U" and were extremely swift runners. Both species were incredibly agile, leaping 20 feet or more between trees with apparent ease. Even mothers would perform these leaps, with infants clinging tenaciously to the fur on their abdomens.

One of the most distinctive sounds was the resonant "whoop-whoop" of the langurs in the early morning, each troop advertising its presence throughout the jungle. Troops vary in size from twelve to thirty or more individuals, the lead males clearly recognizable by their large size. Life in a langur troop seems more peaceful to the observer than in a bonnet macaque troop. In the latter, screams and growls, together with frequent crouching and appeasing grimaces by subordinates, indicate a more rigid social organization based upon a dominance order. Even so, both species engage in long bouts of social grooming, and infant monkeys receive a lot of attention and social freedom compared to adolescents and subordinate adults.

The Indian elephant *(Elephas maximus)* was commonly seen in the study area. Whereas in the adjacent Mudumalai sanctuary family groups were frequent, solitary males predominated along the rivers in the bamboo groves. Occasionally, lone males were found at much higher elevations in the surrounding hills. Many of these solitary males were dangerous and represented a major hazard to our work. They are often shot at by villagers when they break into their fields and ravage their crops, and this treatment made them distrust humans and likely to charge at us.

Family groups consist of one or more adult females and usually two or more juveniles, with perhaps a very small infant never left unattended. When dispersed for feeding, the family members maintain intermittent contact with each other by trumpeting. Two or sometimes more adults and subadults will carefully assist an infant elephant across a river, supporting it with their sides and trunks. Interestingly, elephant society is matriarchal, the solitary males returning to the family herds for brief periods when breeding takes place. Occasionally two males will consort together, but more often they live alone and have little to do with the family group's social life.

On two occasions, elephant "dens" were found in dense scrub near the rivers. One was a single chamber approximately 20 feet across. The entire area was heavily trampled, with dung pushed to the periphery. The bushes had been flattened to form a wall around the "den," which was shaded by a tall and leafy banyan tree. The other side of this tree was reddened with mud and dust from the elephant's body. These animals use favored trees as rubbing posts, and the mud and dust they put onto their backs with their trunks most likely serves as an abrasive for grooming purposes. The second den was similarly located under a large tree and consisted

Figure 3.1 Bonnet macaque mother and infant.

of two chambers of comparable dimensions to the single one described. Such "dens" are probably resting areas habitually used by solitary tuskers. For reasons of personal safety, however, we spent little time in the vicinity of these dens. They were located in dense scrub and accessible only via ele-

Figure 3.2 Mother, infant, and adolescent common or grey langur monkeys.

phant trails, narrow corridors that offered neither escape exits nor cover if one were to meet an elephant in transit!*

With destruction of its habitat, the elephant is being forced to move

---

*A senior Indian government wildlife official was killed by an elephant while walking in the jungle with Johnsingh during the latter's study period in the Bandipur Sanctuary.

Figure 3.3 "People" of the trees: three bonnet macaque monkeys.

Figure 3.4 Solitary male elephants were often encountered, and respectfully avoided.

Figure 3.5 Usually traveling in small groups, sambar does cautiously ford a river.

from areas of agricultural development into more remote regions of the jungle and into protected sanctuary land. Consequently, elephants, the most habitat-destructive of perhaps all mammals when concentrated in large numbers (with the exception of man, of course) are being forced into a shrinking space. Unless their numbers are carefully controlled they will become a major threat to habitat conservation, a consequence of man's agricultural encroachment on the jungle.

Figure 3.6 Part of a herd of forty-five to fifty chital or spotted deer at early morning drinking site on Sigur River.

The sambar *(Cervus unicolor)* is the largest Indian deer, males attaining a shoulder height of nearly five feet and weighing over 700 pounds. The coat is coarse and shaggy and, in stags, forms a distinct mane around the neck and throat. The coat's color is a solid brown with a yellowish or greyish tinge, females lighter than males, older stags becoming very dark. The sambar is an elusive animal, essentially nocturnal and seeking cover in dense brush during the day. In spite of its size, it is a favorite prey of the whistling dog. While young stags remain with the herds, which travel in small numbers of two to five, mature stags live a virtually solitary life except during the rutting season. During this period they tend to be highly territorial, competing for a particular locale and breeding with the hinds that enter their territory. Sambar have a distinctive, resonant "honk" alarm call like a bell or truck horn, and in spite of their size and weight they can move with incredible silence through dense jungle.

The chital or spotted deer *(Axis axis)* is abundant in the study area and was a major prey of the whistling dog. Both fawns and adults have white spots on their fawn-colored coats, which tend to darken with age. The lower series of spots on the flanks are arranged in longitudinal rows and form broken, linear markings. At certain angles the spots cannot be seen, and, at a distance, a small fawn may be easily mistaken for a wild dog, which has similar rust brown coloration, large ears, and a leaping, springing

Figure 3.7 Chital stag and doe, common prey of the whistling dog.

gait in tall grass. The latter, however, has a distinctive dark tail, while the short tail of a deer is clearly recognizable by the white ventral side which, when displayed, is an effective alarm signal. This displayed tail, with the "toot" alarm call, is often elicited by the presence of a predator or human intruder. Daily records of chital alarm calls (see Figure 3.8) disclose a pattern of early morning and evening peaks, correlating with the crepuscular hunting patterns of wild dog and, possibly, of panther, at dusk.

The chital's barklike toots are sometimes repeated several times in succession and, occasionally, one animal will answer another in a duetting cadence. Such repeated sounds, lower in intensity, may be used by deer to locate each other, especially when in dense scrub and assembling to "roost" at night.

The chital is a "brazer," that is, it both browses on shrubby foliage and grazes on grass in open areas. The animals tend to aggregate in large herds of forty to sixty or more for "roosting" at night, usually in open areas. The herd disperses soon after sunrise, breaking into small feeding groups for the day. During the dry season their activities are restricted because water is only available in pools along the drying rivers. Also, since food is scarce during this period, larger aggregations of deer are not often seen.

Chital, therefore, have two seasonal patterns of social behavior. In the dry season, when food is scarce, they graze in small groups. During and soon after the rains, however, food is abundant and they need not disperse

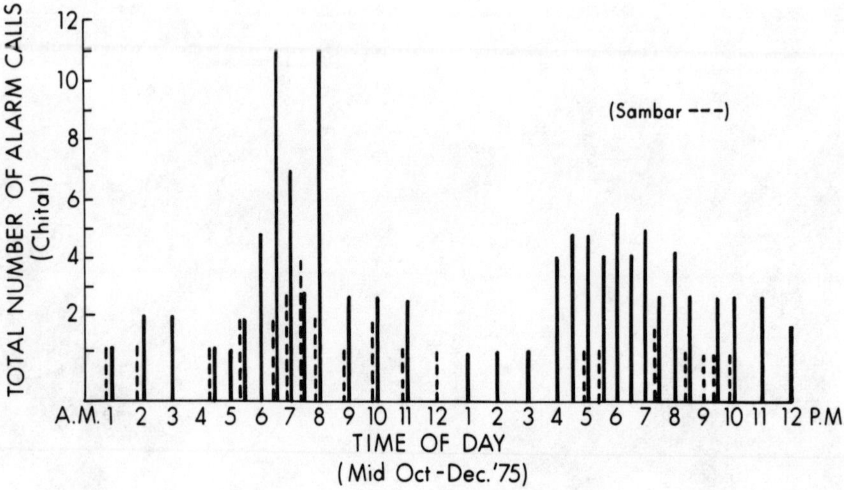

Figure 3.8 The frequency of calling (single or repeated alarm calls) in spotted deer (chital) and sambar coincides with activity peaks of predators.

to find food. Hence, larger herds are to be found in the jungle at this time, especially in their favored ecotone—the borderline between jungle scrub and more open park or meadow. In larger herds, the deer are less vulnerable to predation. Our analysis of wild dog feces confirms this fact; during the October–November wet season, wild dogs were eating less chital than in the dry season. Also more small rodents, black-nape hare, and domestic livestock (buffalo and Brahmini cow) were found in the dogs' scats during the dry season.

Two other species of deer were found in the steep slopes of the Nilgiris close to the study area, the muntjac or barking deer *(Muntiacus muntjak)* and the mouse deer *(Tragulus meminna)*. The former is a chestnut red color. Primarily nocturnal, these shy, solitary deer, rarely over 30 inches high at the shoulder and weighing no more than 50 pounds, are confined to the dense vegetation on steep slopes in the surrounding Nilgiri hills. Occasionally, barking deer are seen in small family groups, males having two bony, hair-covered pedicels out of which a pair of short antlers grow. Females are recognized by tufts of bristly hair which replace the antlers. These pedicels continue as a bony ridge down each side of the face, hence, this small deer is often referred to as the rib-faced deer. When alarmed, barking deer will emit a clear, doglike bark and, when in flight, they may sound a volley of cackling barks.

The mouse deer stands no more than 10–13 inches at the shoulder; it has no antlers, but, instead, uses small tusks for defense. The coat is olive brown, speckled with yellow, with rows of buff or white spots and bands on the flanks. This deer occupies the same habitat as the muntjac, and be-

Figure 3.9 Wild pig: sow and two piglets, part of a small herd of twelve. These animals occasionally fall prey to the wild dogs.

cause of its small size will use rocky crevices as hides to escape predators. According to S. H. Prater's *The Book of Indian Animals* (1965), these deer come out to feed in the mornings or at dusk and never venture far into the open, bolting into a nearby "hide" when alarmed. Apparently, such hides are used by a doe to raise one or two young. Little else is known about these intriguing little ruminants, and in the study area they were rare.

The Indian wild boar *(Sus scrofa)* occasionally falls prey to the wild dogs. This animal is omnivorous, living on roots, tubers, wild fruits, insects, snakes, and even carrion. Extremely alert and difficult to observe, the wild pig is secretive during the day, being most active at night and early morning. The boars tend to be solitary except during the breeding season, but may occasionally be seen with another boar or one or two sows. Quite large herds of twenty to forty sows, piglets, and juvenile animals attest to their high sociability. Their activities are apparent throughout the jungle, where large areas of ground have been rooted up by herds searching for roots, tubers, and other food.

Another digger and rooter is the sloth bear *(Melursus ursinus)* which, to our knowledge, is not prey for the wild dog. These animals are most active at night, their staple diet consisting of termites, insects, and larvae. With powerful front claws they will tear open termite mounds and rotten logs, and roll over quite large boulders. Sloth bears are solitary, cubs re-

Figure 3.10 Otters, jackal, tiger, and gaur inhabit the Nilgiris (tiger in captivity, Delhi Zoo).

maining with the mother until two to three years of age. The cubs are reportedly carried on the mother's back to the feeding grounds. When food is scarce, sloth bears may eat carrion, but normally the supply of fruit, wild honey, and insects is sufficient for their needs. Because of their large size, up to 320 pounds, they have to travel far in search of food each night.

The Indian porcupine *(Hystrix indica)* is another nocturnal creature rarely seen, although its tracks, scats, and quills left along the trails attest to

Figure 3.11 Indian leopard or panther; both black and spotted phases were present in the study area (captive animal, Delhi Zoo).

its abundance in the Nilgiris. The formidable quills of this animal protect it from all predators; white, hollow "rattling quills" on the tail are used as a threat signal. Reportedly, the porcupine has killed panthers by literally running them through with their quills, driven in by running backward into the assailant. Porcupines are vegetarians, eating fruits, roots, and all kinds of vegetables. Littering their burrows (which may be forty to sixty feet in length) are heaps of gnawed bones and antlers brought back from their foragings. These bones provide them with the large quantities of calcium necessary for the growth of their quills.

The Indian pangolin *(Manis crassicaudata)* was one of the most extraordinary residents of our study area. Its entire body and long tail is covered with a plated armor of scales. This nocturnal animal has powerful front claws for digging, an adaptation that secures its staple diet of ants and termites. In searching for tree ants, pangolins become arboreal and they have a strong prehensile tail adapted for this purpose. Like the South American anteater, pangolins possess a long glutinous tongue which they use to pick up termites, ants, and eggs after breaking into the nests. Like the armadillo, when alarmed they will roll themselves into an impenetrable, tight ball. Probably because of this mode of defense, they are easy prey to at least one predator: man. We saw pangolins only once, but the remains of a fire and pangolin scales nearby indicated that these animals are eaten if caught by hill people.

In the rivers, we occasionally saw a family of four to six clawless otters *(Anoyx cinerea)* fishing and playing together. Extremely alert and quick to detect human presence, a short whistle alarm note would send the whole family into hiding. When undisturbed, these exquisite swimmers would dive and chase each other through the water, often continuing the game as a wrestling play-fight on a sand bank or among the rocks at the river's edge. The otters had a wide repertoire of vocal sounds, ranging from a playful squeaky purr to a more protesting yelp-growl and a loud yelp-bark that was a clear indication of intense excitement. One otter, even though almost fully grown, was seen shining and squeaking as it tried unsuccessfully to solicit food from one of its parents who sat in the shallows crunching a freshly caught fish. A pair of much larger common otters *(Lutra lutra)* was also seen fishing and playing together in the dammed lake at the study area's western edge.

In the surrounding steep hills, small herds of gaur or Indian bison *(Bos gaurus)* were sometimes found. Favorite trophy animals for the hunter, these magnificent creatures are making a good recovery following their decimation a few years ago by an outbreak of rinderpest. A mature bull may stand almost seven feet at the shoulder, with a horn spread often attaining three feet. With increasing maturity, the light, golden yellow coat of the calf turns to fawn, and then to a reddish brown or coffee color in cows and young bulls; adult bulls are black.

Although adult bulls may attack man without provocation, they are generally timid and therefore difficult to approach and observe closely. The bulls do not apparently hold territory nor exclusive rights over a herd of cows, and tend to roam either alone or in the company of other mature males. Some do roam more extensively during the breeding season in search of herds, but they do not stray far from the main herd. Gaur generally live in small family herds of eight to twelve animals, occasionally uniting with other families in large pasturing areas.

This magnificent creature has an optimistic future. It is more likely to fall prey to the tiger, rare in the Nilgiris, than to wild dogs, and its chances of contracting rinderpest are diminished since domestic cattle are now vaccinated against rinderpest and hoof and mouth disease. The gaur's improved status is also enhanced by being the most popular trophy animal for the hunter—perhaps the cause of the gaur's existence today, since, in the past, there were no other motives to preserve it.

The leopard or panther *(Panthera pardus)* was more active in the study area during our second visit, possibly because of more rigorous controls over the use of poisons by farmers to kill predators. One leopard was observed in midafternoon, resting in the shade, with domestic buffalo close by and quite unconcerned about its presence. One leopard kill was found, a

Figure 3.12 Evidence of leopard activity: a five-month-old Brahmini steer carcass cached in a tall tree where it is safe from hyenas and jackals.

five- to six-month-old Brahmini calf which had been characteristically dragged into a tall tree, safe from wild dogs and hyenas. Mr. G. J. Rajasingh, chief conservator of Tamil Nadu Forest Reserves, once observed two wild dogs alert and waiting for a leopard to finish eating its kill. Instead of eating the big cat's leftovers, the wild dogs gave chase to the leopard! Although agile and powerful, the leopard is obviously wary of wild dogs, whose coordinated attacks could lead to its death. Leopards are generally solitary nocturnal hunters, but they may also hunt during the day in cooler months. Their prey ranges from sambar, chital, and domestic livestock to birds, reptiles, and, occasionally, monkeys and porcupine.

Chapter 4

THE WHISTLING DOG

The whistling dog is found in central and east Asia, from China (Manchuria) in the north to India and Malaya in the south. According to Prater (1965), they are also found on the islands of Sumatra and Java, but are absent in Borneo, Ceylon, and Japan. They occupy very different habitats, ranging from the open steppes of Tibet to the scrubland and forests of the south. In the USSR, the dhole lives in dense forests and mountainous alpine regions such as the Ural mountains (Novikov 1956, Sosnovskii 1967), further evidence of their adaptability.

The dhole has been placed in the subfamily *Simocyoninae* of the family *Canidae,* along with the South American bush dog *(Speothos venaticus)* and the African or Cape hunting dog *(Lycaon pictus)*. This taxonomic association has been questioned by Kleiman (1967) and Fox (1971). Clutton-Brock *et al.* (1976), after analyzing thirty-seven canid species on ninety ecological, behavioral, and morphological characteristics, conclude that the dhole is more similar to *Canis* and *Dusicyon* (a coyotelike South American canid, see Fox 1975) than to either *Lycaon* or *Speothos.*

Cuon alpinus is officially regarded an endangered species by the United States Department of the Interior (1974 List of Endangered Fauna), and two subspecies, the Eastern Asiatic wild dog *(Cuon alpinus alpinus)* and a smaller Western Asiatic wild dog *(Cuon alpinus hesperius)*, are recognized as endangered by the International Union for the Conservation of Nature *(1969 Red Data Book)*.

Thenius (1954) has discussed the possible origins of the wild dog using fossil remains. Teilhard de Chardin (1970) described the earliest remains of this animal from the early Pleistocene era in China. It was also widespread during the later part of this era throughout Europe. Thenius believes the wild dog did not exist before the Pleistocene and that it may have evolved from a primitive, jackallike canid. Fossil forms from the late Pleistocene show climatic differences, larger wild dogs occurring in the north (Siberia) and slighter ones in the south (Mediterranean), a phenomenon referred to as Bergmann's rule. Thenius regards the Cape hunting dog and whistling dog to be "catlike dogs" because of their long, fine limbs and backbones. The Cape hunting dog is thought to have evolved from the *Canis* stem leading to present-day jackals.

Whistling dogs' coat color may vary from rufous grey to brown grey and mahogany red; black-tipped hairs may extend along the back and also on the ears. Pale, sandy-colored dogs were seen in the study area in the Nilgiris. The bushy tail is usually tipped with black, more rarely with white. A white or cream-colored throat patch is common and this color may extend under the chest and belly. Cubs are a uniform dark slate gray or brown when young, a characteristic of all wild canids (Fox 1971).

Regional variations in size are evident, the wild dogs of Malaya being much heavier in build than the subspecies studied in southwestern India. Adult males may reach over 40 pounds in body weight (averaging 10 pounds heavier than females), standing 17–22 inches at the shoulder; body length is about 3 feet and the tail 16–17 inches long. Females have twelve to fourteen teats instead of the usual ten for *Canis* species. A unique feature of this canid is that there are six instead of the usual seven molar teeth in the lower jaw.

As in the Cape hunting dog of Africa, the erect ears of the whistling dog are rounded rather than pointed as in the wolf. Males may develop thick cheek tufts, a secondary sexual characteristic which may have some social significance in terms of dominance and social organization.

Because wild dogs are generally so uniformly marked and there is usually little color variation between members of a pack or clan, individual recognition is extremely difficult for the field observer. Unlike the Cape hunting dog, the males do not have a clearly visible prepuce and it is difficult to distinguish between the sexes even at close proximity. These factors alone make it virtually impossible to conduct a detailed field study of social relationships within a pack of twenty or more animals. Without trapping and marking each animal to enable individual recognition at a distance, many questions regarding their social life will remain unanswered. It is not yet known if there is a dominance hierarchy in the pack, although a leader may be recognized. Evidence of a leader in the pack, who engaged in raised-leg

Figure 4.1 Unity and uniformity—almost invisible against any background, a pack of nine whistling dogs. (Photo courtesy of E. R. C. Davidar)

urination and was the focus for displays of active submission by other adults, has been given by Johnsingh (1982). It may also be deduced from his observations that dominance among females results in only one pregnancy and litter.

At times we really believed the wild dogs were deliberately avoiding us. While we worked in one area, they would be reported in another location two or three miles away. We would then go there, find no fresh signs, and return to the former place only to discover they had been there in our absence. To attribute a conscious intent on their part would naturally be going too far, but the idea was not unwarranted considering the state of mind that developed after continuous weeks of frustrating attempts to observe these elusive and wholly unpredictable creatures.

The wild dogs' hunting strategy clearly involves a highly randomized and unpredictable pattern that does not correlate with the movement of deer, since the latter concentrate in one area for some time. The presence of deer did not necessarily ensure that we would see the wild dogs. We consistently worked in one area for five to seven days and then moved to another area in the dogs' hunting range. From this work pattern, we gained an impression that the dogs tended to concentrate in one area for two or three days and did not return to it again for ten to fourteen days or longer. During, and shortly after, the rains their hunting range was more extensive since water holes located several miles from the two rivers (Moyar and Sigur) provided them and their prey with water. In the dry season, the activities of prey and predator alike were more confined to places along the rivers where water was still available in stagnant pools. The peak of the dry season (between March to early May) was therefore the optimal time for study, when the ranges of wild dogs and their prey are very much restricted by the availability of water and vegetation for the deer.

In the first study, at the beginning of the dry season (January to March), visibility was enhanced by the defoliation of deciduous scrub, and more kills were located along the river water pools than in the second study. Even during the wet season, though, the deer tended to run downhill or toward water when pursued, or they were ambushed by the dogs while drinking at the river. The second study coincided with the onset of the denning season (between November and January). The inaccessibility of many dens, however, with the difficulties in locating active dens and the near impossibility of approaching an active den without being detected, made studying the wild dog a challenging, if not overambitious, endeavor.

The Nilgiri valley is an ideal habitat for wild dogs, providing ample cover and many different sites for denning. Dens were extremely difficult to find, but with luck and many hours of following wild dog tracks (aided by Sikka Bunda, an experienced shikari who knew the area well), we learned much about the animals' denning habits. We studied two den sites in active use, three recently abandoned, and two not in use. Two recently abandoned dens were heavily infested with fleas, probably the reason the families had relocated to clean den sites within a mile's distance. Both active and recently abandoned dens were otherwise remarkably clean—we detected no odor of dogs or carrion either inside the dens or in their immediate vicinity. No adult feces were present within 200 yards of any of these dens, and only an occasional cub dropping was found; the adults presumably ingested most of the young's excrement, a habit characteristic of canids. Another reason for leaving the dens may have been to move the newly weaned cubs closer to water, since both abandoned dens were some distance from the nearest water hole.

The various types of dens encountered were classified as follows: (1) simple earth den with one entrance; (2) complex cavernous earth den with more than one entrance, sometimes more than one den in the same area; (3) simple cavernous den, excavated under or between rocks, with one entrance; and (4) complex cavernous denning area, with several dens in the same vicinity, some of which may interconnect. From the numerous bones deeply buried at the den's entrance, we deduced that type 1 may have been "remodeled" hyena or porcupine dens. The striped hyena likes to live in a "graveyard" and porcupines collect bones, an essential part of their diet, as discussed earlier.

Both the simple and complex earth dens may be located on the plains

Figure 4.2 A rocky knoll (a) is a prime site for wild dog dens. In the "playground" of this site (b), pelvis and lumbar vertebrae of a chital kill indicated that adults may carry food to the young. Dens may be dug in the compact earth in the side of a creek, or nullah (c). Near one such den site, the skull of a leopard (d) lies in a clearing worn down by the feet of wild dogs and their cubs.

under dense scrub or on the banks of a river or dry creek (nullah). In the former locale, the entrance may be almost vertical, with a sharp turn located three or more feet down. The den entrance may be large enough for a human to enter, but very often there is a tight squeeze at the turn, thus preventing entry into all but the first 4–8 feet of the tunnel. If this constriction can be negotiated, or if it is absent, the entrance tunnel will open into an antechamber from which one or more passages may extend. One complex earth den, excavated into the bank of a nullah, had six entrances leading to a labyrinth of at least 100 feet of interconnecting tunnels (see Figure 4.3).

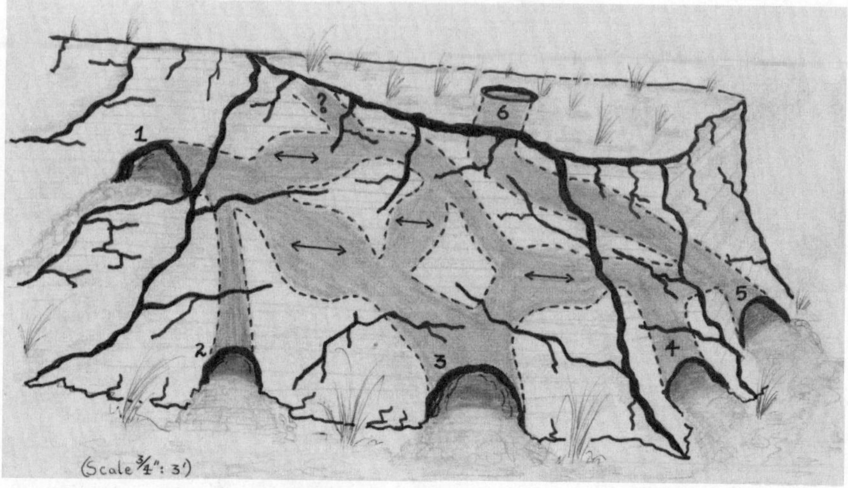

Figure 4.3 Field sketch of a complex den with six entrances dug into the bank of a creek or nullah. Arrows indicate level chambers which are linked by interconnecting passages.

Four large chambers were found inside, and countless hungry fleas.* Many generations of wild dogs must have developed this elaborate "city" and bitches shared the labyrinth and raised their litters together. This protocultural phenomenon, where generation after generation use the same traditional denning sites, is something to respect and cherish. If the wild dogs are to survive, such areas should certainly be kept off limits to trapping, poisoning, and hunting, and every measure should be taken to conserve them. Denning time is the dogs' most critical and vulnerable period.

Close to this complex den was a small earth den with two entrances either dug up by a nonconformist or excavated in a year when several bitches were in whelp and there was not sufficient room in the communal "city" for all of them. Across the nullah was an open area, and under shade trees we found four places where the grass was flattened down, each 8–10 feet in diameter. A few bones and skin of chital, sambar, and wild pig, and the skull, humerus, and scapula of a leopard were lying in these little clearings. This area was used by the cubs for playing before the pack moved to a new location.

Similar play areas were found in two active denning areas, but bones were present at only one of these. An intriguing series of short tunnels was formed in one of these playgrounds in a stand of long grass that had not been trampled down by the dogs. This clump of grass was used by the dogs for ambush-play. We saw three adult dogs early one morning using this

*Later identified as *Ctenocephalides felis felis*.

Figure 4.4 A long and deep rocky crevice provides shade for several members of the pack that occupy this den site, which includes the usual playground. The site is, however, unique because of the presence of grass tunnels used in ambush games.

"structure" for their games (discussed later). Close by was a fresh spring, as well as a magnificent view of the valley below, and the remains of a wild pig the dogs had enjoyed for breakfast that morning. Paradise indeed!

This latter denning area belonged to the hill pack that hunted each

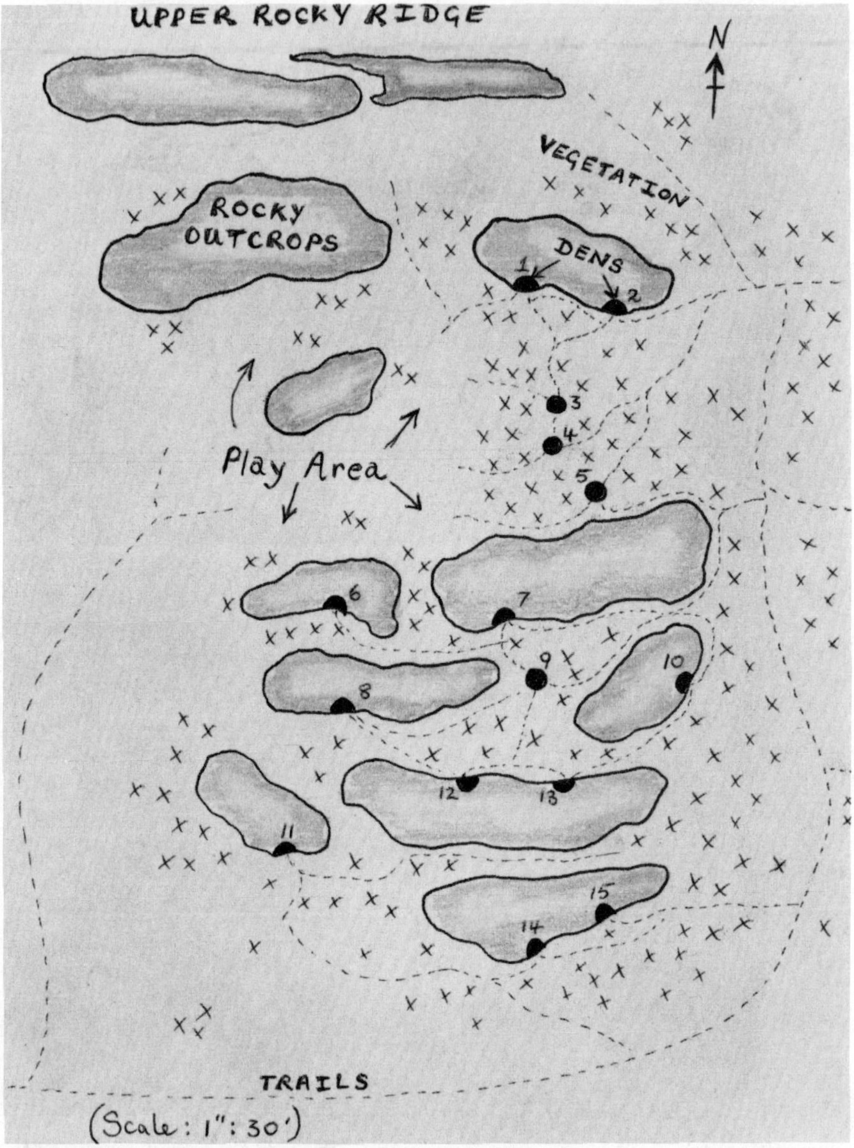

Figure 4.5 Field sketch of multiple dens under large boulders on a rocky knoll. Note that most trails lead out and in from the east. The prevailing wind here comes from the east.

day in the valley below. Four simple cavernous dens were found at this site, and a fifth, 100 yards farther down the hill, included a play area in front of it.

The other active denning area belonged to a pack that lived 8 miles northeast of the two packs we were following. It was located along the valley in a rocky ridge that rose some 100 feet above the plateau. No less than twelve cavernous dens were found in this "city," plus three additional earth dens, some of which were interconnected (see Figure 4.5). Most of the tracks to and from this "city" were on the east side, reflecting the whistling dogs' habit of moving into the wind when hunting so that prey is easily scented and the scent of the hunters not detected. The prevailing wind was invariably from the east, according to our shikari.

A porcupine was in residence in one of the simple cavernous dens of this area, and two mongoose resided close by, coming out regularly between 7:30 and 8:00 A.M. and near 4:00 P.M. to hunt for food. A study of the relationships between the wild dogs and other animals which are potential prey and which share their denning area would be interesting. Elsewhere, red foxes and dingos have been known to ignore ducks and rabbits residing within their den sites.

Although a cub was seen at this denning site, no cub nor adult feces were found and all dens examined were odorless and free of fleas. Between our visits to this "city," the pack had brought a chital's lumbar and sacral vertebrae with pelvis still attached, presumably for the cubs. These bones, together with the tracks of several cubs and small holes they had been digging, were found in the playground area (see Figure 4.2). Johnsingh (1982) also found play and rest areas 2–3 square meters in diameter in shaded spots near all the dens in his study area, but did not find any bones or skin nearby. He attributes the presence of bones in the dogs' denning areas to hyenas (along with many fleas—the four occupied dhole dens he studied were free of both fleas and odor).

These findings reflect the highly evolved family life and social structure of the whistling dogs. However, since the animals are so elusive and difficult to identify individually, our knowledge is limited. What is known, and what we were able to piece together in our study of the Nilgiri chennai, brings us even closer to these remarkable hunters.

Chapter 5

THE WHISTLING HUNTER

Wild dogs usually hunt in the early hours of the morning, between 5:00 and 8:30 A.M. They generally avoid strenuous activity during the heat of the day, seeking shade under rocks and dense underbrush or lying in or along the banks of a river. This behavioral thermoregulation greatly influences the daily activity and movements of these dogs living in tropical and subtropical regions, especially in the hot, dry season. In cooler, rainy seasons, they may be active and hunt any time of the day. They rarely hunt at night, but are often active on moonlit nights. This fact implies that they rely greatly on the sense of sight for hunting; but, in fact, all senses are used flexibly without any particular specialization in any one modality. Sound and smell is of value when prey "freezes" and is concealed in dense cover, and sight and smell is utilized when tracking across more open country.

Fresh kills can be found by "reading the sky" from a high vantage point, that is, heading toward a group of circling or roosting vultures. This method can involve a mile or more of looking through dense, thorny scrub, only to find the vultures were simply circling (for enjoyment?) on a hot air draft or congregating in a tree to bask in the early morning sun. Another, just as reliable, way to find fresh kills is to look for a jungle crow cawing from a tall tree. If it flies toward another crow calling in the distance, there is a chance the latter bird has found a kill or is following a pack of wild

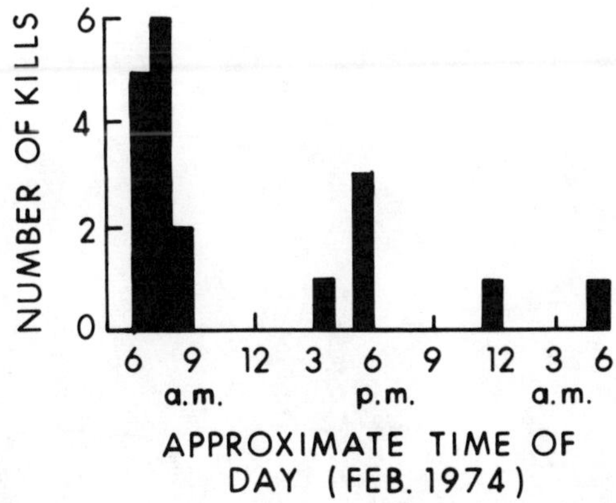

Figure 5.1 Whistling dogs most often hunt and kill between 6:00 and 8:00 A.M., thus avoiding the heat of the day (data from direct observations, finding fresh kills, and reports from graziers).

dogs. Sometimes the crow will simply be cawing at a sambar under a tree or at a small group of browsing chital, or just cawing to its companion. If we had known the language of the crows better, we might have saved hours of unproductive walking, but, even as it was, the crows did help us find both kills and wild dogs. A combination of vultures and one or more crows calling in the same tree is the most positive sign of wild dog activity.

Interestingly, dholes were observed on three separate occasions by Johnsingh (1982) allowing wild pigs to scavenge the remains of their kills. They would chase away village dogs, but, on one occasion, did not harm

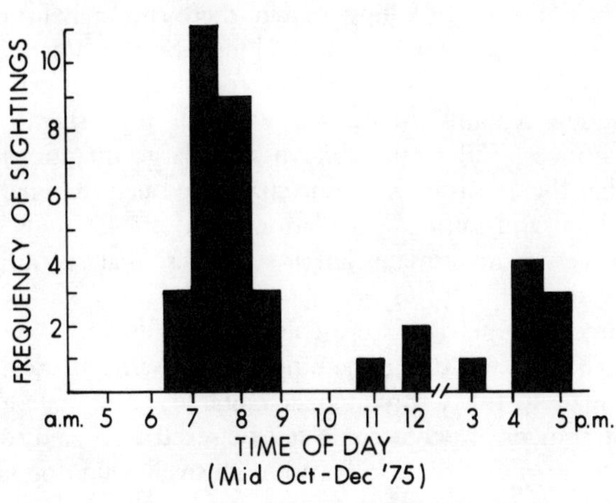

Figure 5.2 Whistling dogs were most often seen between 6:30 and 9:00 A.M. and 4:00-5:00 P.M., at which times they were most likely to make a kill.

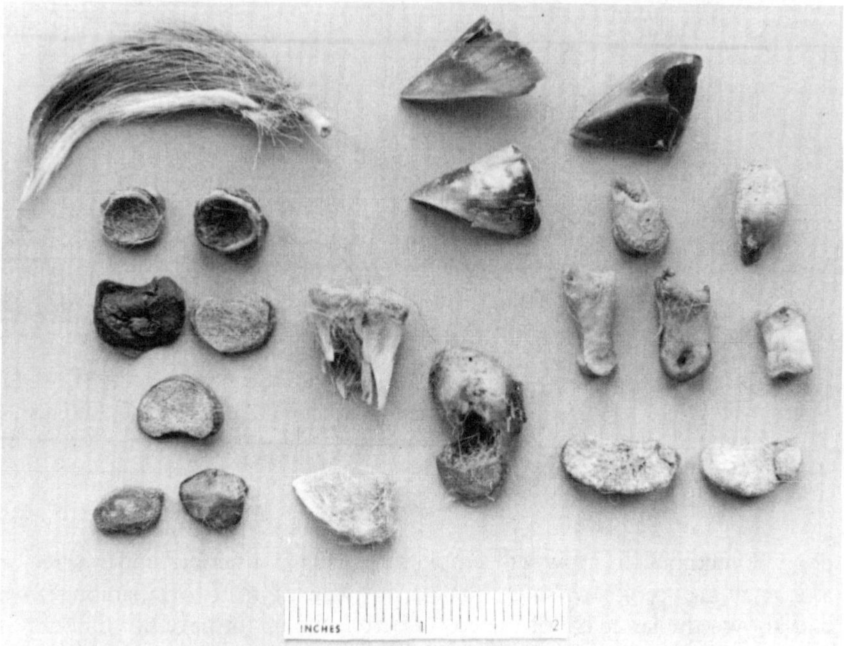

Figure 5.3 Chital fawn remains found in wild dog feces include fragments of bone, toe horn, tarsal and carpal bones, intervertebral discs, and epiphyses (ends of long bones). A whole tail was also found (upper left)—probably pulled off and swallowed by the dog that seized the fawn by its tail during the kill.

their pups which could have been easily killed as they lagged behind their mother.*

Within half an hour of killing a fawn, there is usually little left. Shreds of the prey's skin, an occasional piece of long bone such as a femur, and a few fragments of skull may be found before vultures and other scavengers clean up these few remains. Analysis of wild dog feces shows the presence of fur, skin, hooves, and teeth of fawns and large quantities of digested bone, which has the consistency of fine chalk. To calculate what percentages of the total chital and sambar population killed are fawns is difficult, because there are few if any remains left at such kills; analysis of fecal contents is, therefore, essential.

Sixty-three deer pelvises were collected in the study area, fifty-six of which were chital, and the remaining seven, sambar. Sex ratios were twenty-three male to thirty-three female chital and one male to five female sambar (one fragment of pelvis could not be sexed). To gain some more insight into the maturity of these kills (not exclusively wild dog kills), an arbi-

*Johnsingh also observed elephants chasing the dholes on several occasions.

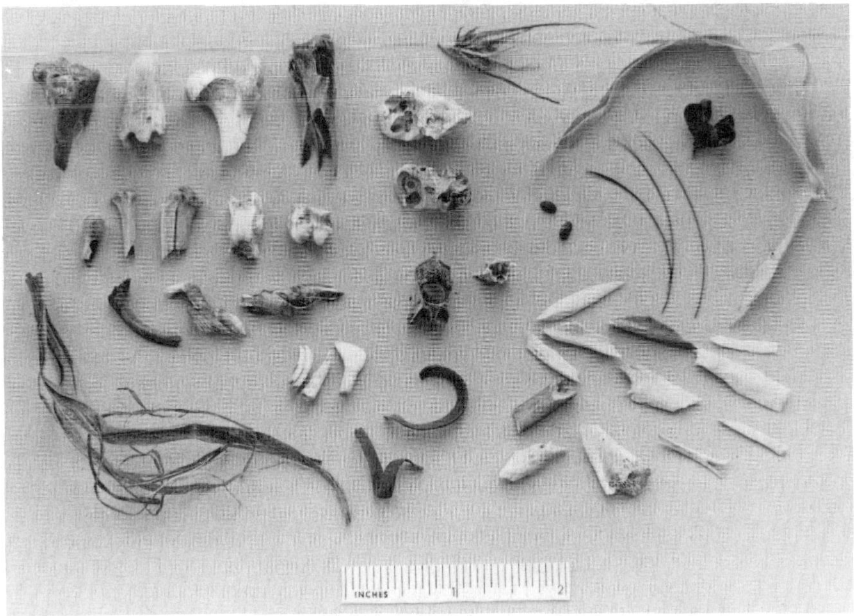

Figure 5.4 Miscellaneous items in wild dog feces include grass and grass seeds, wild boar bristles, teeth of chital fawn, tendons, eye lenses, bone fragments, and claws of hare and mongoose.

trary pelvic growth-size index was taken by measuring the distance from the top (anterior rim) of the acetabulum, or hip joint, to the bottom of the obturator foramen. This index was chosen since it was the most intact region—the ilium and ischium usually being chewed and splintered by the dogs.

According to Figure 5.5, pelvises with an index less than 7.5 centimeters must have been ingested entirely by the wild dogs, since no remains were found in the field (with the exception of a two- to three-month-old fawn killed by three dogs). A pelvis with a 7.5-centimeter index would be well ossified, especially at the region of the acetabulum, and could not, therefore, be easily ingested. From the extrapolated graph it could also be concluded that wild dogs do not kill many fawns or subadults with a pelvic index of 6.5–7.5 centimeters, but analysis of fecal contents indicated the contrary. The absence of pelvic remains of subadult deer in the spring collection of this study is an open question. We do not know how long moderately calcified bones remain intact. Their half-life may be as short as two to three months, since many scavengers, notably porcupines and small rodents, will ingest small bones that have a rich source of mineral salts essential to the animals' diet. It may be warranted, to conclude, since no subadult pelvises were found, that subadult fawns are more difficult for the

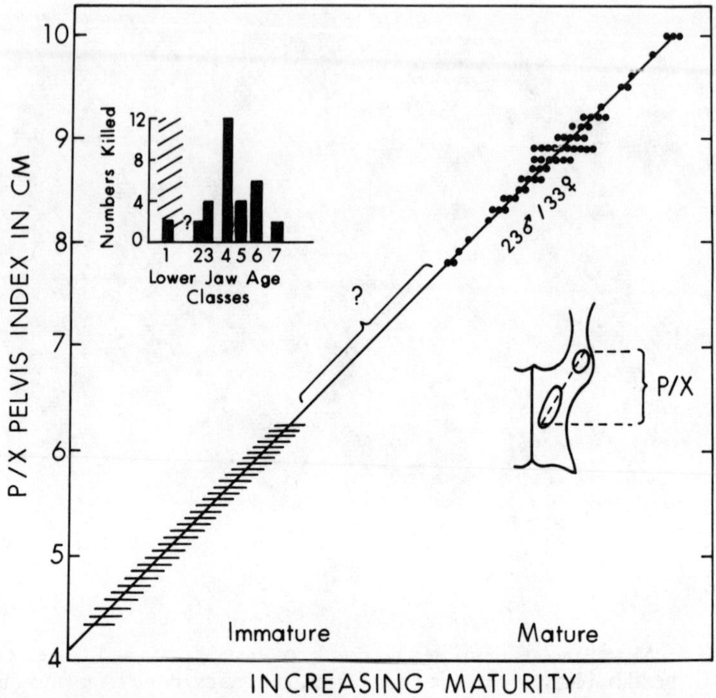

Figure 5.5 Graph of pelvis size indicates that juveniles and large males are rarely taken. Shaded area indicates projected scale of pelvic index in immature deer, no samples being found in the field. Most specimens found were fully mature. More accurate age estimates are afforded by aging the teeth: the greatest number of kills being in the "prime" categories three-six (see text). Most fawn jaws were eaten and few were found in the field.

dholes to catch than pregnant does, bucks preoccupied with rut, and adults with large and sensitive antlers in velvet.

A similar age gap was found in the samples of lower jaws collected from the same study area (see Figure 5.5). The same arguments posed above hold for the absence of subadult jaws in the collection. Interestingly, the greatest proportion of kills was in the *four to six* year (prime) age group. This observation may not indicate that wild dogs selectively kill animals of this age, but, rather, that the majority of animals in the herds (with the exception of the large annual fawn crop) falls within this age group. This conclusion is supported indirectly by the Nilgiri Game Association's records of the numbers of chital shot each year. The numbers have increased greatly over the last few years, indicating there must be more individuals in the herds falling into this highly productive age range. The age classes of kills identified from lower jaws are shown in Figure 5.5, following Schaller's (1967) age classification based upon the wear on various teeth. Class 1 represents the fawns, and in most of these kills the jaws are eaten by the dogs,

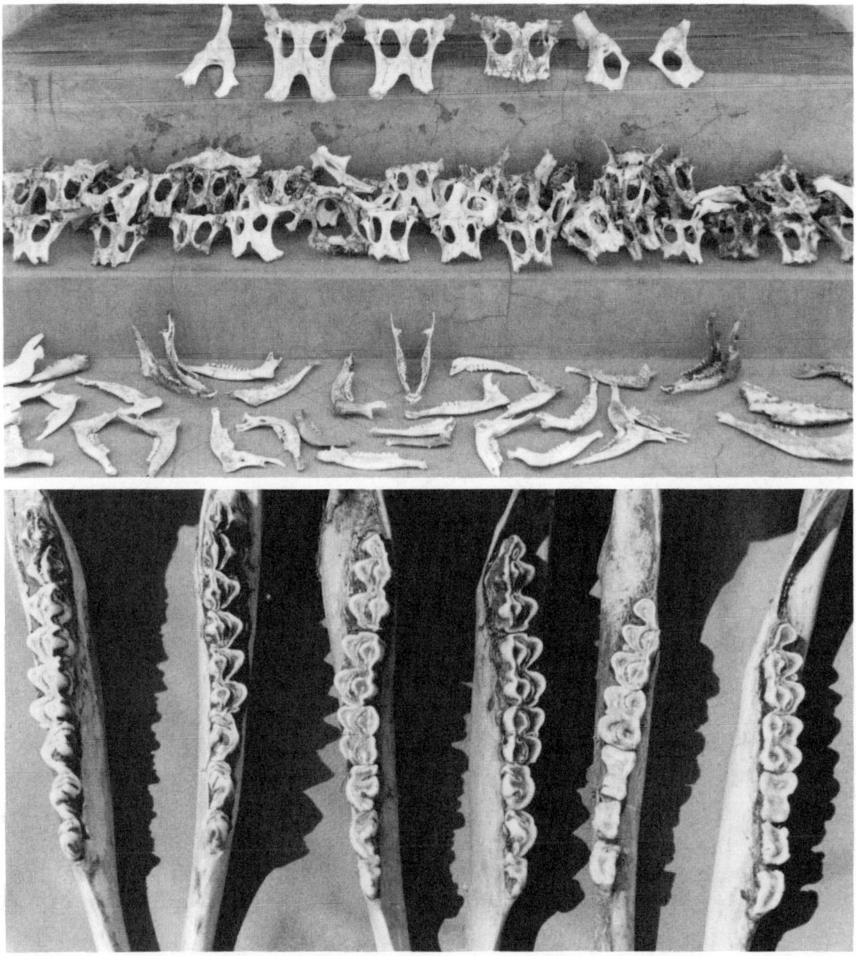

Figure 5.6 Some of the chewed pelvises and lower jaws of chital and sambar killed by whistling dogs. Note the loss of "crowns" and the smoothing of the molar teeth with increasing age (from left to right) in the chital jaws.

since we only found two specimens. Classes 2 and 3 are yearlings and young adults, and Classes 4, 5, and 6 represent prime adults. Class 7 is past prime; interestingly, no really aged specimens were found, indicating that chital in the Nilgiris rarely live over eight or ten years of age. Only eight lower jaws of sambar were collected in contrast to thirty-one lower jaws of chital, and all sambar jaws were subadults.

Analysis of feces (see Table 5.1) confirmed that chital was the most commonly eaten prey during the dry season in our first field study. We were wary about collecting too many fecal samples from the packs since the feces were deposited mainly at communal dunging areas (see Chapter 6)

Table 5.1 *Fecal Analysis*

	(138 samples)				
	Chital	*Sambar*	*Domestic Cattle*	*Wild Pig*	*Small Mammals*
Adults	102	13	3	1	12
Pups	5	0	0	0	2

Iseilema prostratum grass common in feces, especially with sambar remains.
Zyzyphus fruit present in one sample.
Segments of tapeworm *Taenia hydatigena* were frequently found.

and their collection might have had some effect on the dogs' marking behavior.

The ratio of chital to sambar remains in the feces was very close to the ratio of chital to sambar pelvises collected in the field (approximately 10.0:1.5, or twenty chital killed for every three sambar). This ratio accords with the lower population of sambar and with the fact that these deer are very large (up to 1000 pounds), powerful, and difficult prey for the wild dogs to bring down. Presumably, the wild dogs kill mainly fawns and subadult sambar, since no adult lower jaws were found in the study area.

With greater expertise, we might have been able to distinguish between the fur of chital fawns, subadults, and adults. Judging from the number of feces containing digested calcium (from the bones of ingested fawns) the ratio of fawns to adults during the period December–February was about 2:1. A complete tail of a fawn, many small tarsal and carpal bones, and undigested hooves were commonly found in these feces. Fecal samples collected during the second study period indicated a heavier predation on small mammals, especially hare. Toward the end of the study, however, more fawn remains were found, correlating clearly with the onset of the main fawning season in late December.

Vast quantities of fur were present in many of the fecal samples in association with bone fragments, especially from fawns and small mammals (principally hare and, occasionally, palm squirrel and field rat; one mongoose was also identified). It would be highly adaptive for the wild dogs to ingest fur to protect the alimentary tract from possible injury, especially from the extremely sharp bone spicules from small mammals (see Figures 5.3 and 5.4). Removing the matted fur that formed tight wads and balls around these splinters of bone was very difficult. Clearly, if carnivores in captivity are fed small mammals such as rabbits, the carcass should not be skinned but left intact so that ingested fur may form a protective bolus around any sharp fragments of ingested bone.

The composition of successive stools passed by one wild dog reflected even more of the animals' eating habits. Some stools contained little or no hair, but mainly dark digested meat protein and occasional strips of partly digested muscle and tendon. Other stools contained fur and a lower proportion of digested protein and bones. Bone fragments were never found in stools without hair. The dogs either purposefully ingest quantities of hair when crushing and swallowing bones, or in the process of digestion fur aggregates around bone fragments and both are usually voided together, separate from the meat portion of the meal. Fragments of skull, teeth, claws, and ocular lenses embedded in fur showed that small mammals were eaten whole.

A high proportion of sambar remains included ingested grass *(Iseilema prostratum)*. One fecal sample contained only grass and twigs: this dog was probably sick since the stool was liquid. The only other plant material ingested in any significant amount was the fruit *Zizyphus,* present in several samples. This fruit is also commonly eaten by langur, macaque, porcupine, bear, and man, and is reportedly rich in vitamin C.

Some fragments of grass, seeds, twigs, and bamboo leaves were found in many feces, but their presence was probably accidental—a wild dog eating its prey on the ground is bound to pick up such materials. Longer strands of grass stems, leaves, and flower heads identified as *Iseilema prostratum* must have been purposely ingested by the dogs because they were not fragmented and often occurred in considerable quantity.

There is no satisfactory answer as to why carnivores eat grass. Like fur, it may be an anti-irritant. The presence of grass in sambar remains might have been to protect the alimentary tract from bone fragments. Sambar hair is springy and does not ball and mat so effectively as the fur of chital and smaller mammals. Grass may also be an important source of vitamins and trace elements not available in the all-meat diet, since the stomach contents of chital and sambar were not eaten (discussed later). We were not able to ascertain whether or not the whistling dogs eviscerated small mammals before eating them, but this was certainly the case with deer.

A different predation pattern associated with the postmonsoon period was observed in the second field study, between October and December. A more accurate identification of fecal hair samples (using a wax impression method to reveal the cuticular hair scale pattern of various species) enabled us to broaden the qualitative analysis, although the basic visual identification of sambar, chital, domestic cattle, and small rodents (used in the first study) compared reliably with the analysis shown in Table 5.2. In this postmonsoon October–December period, the wild dogs' diet consisted mainly of hare, with chital, sambar, field rat, and wild pigs present in a decreasing order of frequency.

The relatively high percentage of sambar in the diet may indicate prey selectivity on the part of the wild dogs. A comparison of the total numbers of chital and sambar observed in the field indicates an approximate 6:1

Table 5.2 *Frequency and Percent of Occurrence of Various Prey Species Identified* in 150 Dhole Scats (October-December, 1975).*

Species	*Frequency of Occurrence*	*Percent of Occurrence*
Common Hare	36	24.00
Chital	28	18.67
Sambar	22	14.67
Field Rat	21	14.00
Wild Pig	17	11.33
Domestic Cattle	3	2.00
Muntjac	3	2.00
Mouse-Deer	2	1.33
Gaur	1	0.67
Insects	7	4.67
Misc. (lizards, scales, berries)	2	1.33

*Using Weingart's (Weingart E. L. 1973. A simple technique for revealing hair scale patterns. *American Midland Natur.* 90:508–509) guard hair cuticular scale pattern method from a key made from hair samples of Indian mammals taken from skins at the Bombay Natural History Society, by J. A. Cohen. (See Cohen, J. A., Fox M. W.; Johnsingh A. J. T.; and Barnett B. D. 1977. Food habits of the dhole in south India. *Journal of Wildlife Management* 42:933–936.)

abundance ratio for these species, respectively. The total number of chital, however, is probably somewhat inflated due to multiple sightings of the same individuals. Thus, assuming every chital was counted twice, this ratio becomes 3:1. If the dholes were feeding on a strictly nonselective basis of random encounter, you would expect a similar ratio of occurrence for these two prey species in the dhole scats. The observed ratio, however, was 1.27:1.00. This ratio indicates the dholes were feeding less on chital than would be expected by random encounter and that some form of prey selection must have occurred.

A survey of the local farms *(pattis)* revealed a total cattle population in the study area of approximately 2,650 head. Graziers estimated a 10 percent mortality rate attributable to wild dog predation. Since all these cattle were allowed to roam freely in the jungle adjacent to the patti settlements during the day, the unexpectedly low (3 percent) incidence of domestic cattle in the dogs' diet deserves scrutiny. The obvious abundance of this food source and its ease of capture and killing may be offset by one or a combination of the following factors: (1) the dholes avoid proximity to pattis since they are afraid of people (who will even drive them from a sambar or chital kill to obtain the meat for themselves, especially if the kill is made near a patti); (2) the cattle are let out to graze around 7:00–8:00 A.M. and are corralled at 4:00–5:00 P.M., so grazing periods do not coincide with the early and late day hunting peaks of the wild dogs; and (3) the dhole prefers the harder-to-get but perhaps more palatable flesh of wild prey.

Prey selection may occur in various ways. The activity peak of the dholes corresponds more closely to that of the sambar than to that of the chital. In fact, the latter species was most active during the hour *prior* to the predator's activity peak and only half that active *during* the peak. This reduction in activity was not apparent for the sambar and suggests different temporal availability of these two prey species. These trends are generally confirmed by the alarm-call data.

Furthermore, because animals in large herds are more likely to become aware of a predator's presence than are solitary individuals, the greater social tendency of the chital could also have led to its unexpectedly low frequency of occurrence in the scats. To some extent, however, this advantage would be compensated by the larger size and weight of sambar (4:1 weight ratio; Prater 1965) which makes the latter a formidable prey.

Mech (1970, p. 41) lists four factors which may effect the pack size of wolves:

1. The smallest number of wolves required to locate and kill prey safely and efficiently
2. The largest number that could feed effectively on a particular prey individual
3. The number of other pack members with which each wolf could form social bonds (social attachment factor)
4. The amount of social competition that each pack member could accept (social competition factor)

The same factors may, to some degree, regulate pack size in the dhole. Since the dogs do operate in packs and may kill large ungulates, the first factor probably operates to some extent. The second factor may limit pack size in certain cases, but not always. The stomach capacity of an average 90-pound wolf is about 17 pounds (Mech 1970). If you assume an average 35-pound dhole (Prater 1965) has a stomach capacity to body weight ratio proportional to that of a wolf, then each dhole could consume about 6.5 pounds of prey on an empty stomach. The average weights of adult sambar and chital are 600 and 150 pounds, respectively (Prater 1965). If we further assume that only 50 percent of this prey weight is edible, then as many as forty-five dholes could feed on a single sambar kill, and eleven could feed on a chital. Since the average pack size observed in this study was only five individuals, some other factor(s) must keep the pack size below these maxima. Packs of up to forty individuals have been seen (Davidar 1975). This fact may indicate that, at times, predation patterns may be altered so that the "maximum feeders" factor becomes important. In fact, a pack of forty corresponds closely to the theoretical forty-five dhole maximum for sambar derived above, suggesting that packs of this size may feed primarily upon

Table 5.3 *Frequency of Occurrence of Food Items in 509 Dhole Droppings Collected in Bandipur*

Food Item	*Number of scats with the remains of*	*Percent*
Chital Adult	126	24.75
Chital Fawn	138	27.11
Hare	71	13.94
Sambar Adult	16	3.14
Sambar Fawn	54	10.60
Microtine Rodents	31	6.09
Cattle	14	2.75
Barking Deer	8	1.57
Langur	6	1.17
Wild Pig	4	0.78
Gaur Calf	1	0.19
Beetle	5	0.98
Grass	35	6.87
Miscellaneous (lizards, berries, etc.)	..	..
No identifiable remains	..	..
Total	509	

Source: Johnsingh 1979.

such large prey. The eleven dhole maximum for chital falls into the range of five to twelve cited by Davidar (1975) as the "common" dhole pack size. The social attachment and competition factors listed by Mech may also operate, but no information is yet available to support such a conclusion.

The infrequent sightings of large dhole packs correlate with the rapid decline of the study area's sambar population in recent years. Earlier field observations by hunters and naturalists recount numerous sightings of large packs (within the forty range optimal for sambar). A decline in such observed numbers may indicate a shift in hunting strategy into smaller packs adapted to securing smaller prey species, although the basic preference for sambar, when available, may be implied from the scat analysis.

Table 5.3 summarizes the scat analysis of dholes in the Bandipur study area from Johnsingh's thesis (1979). These data again emphasize that the wild dog preys principally upon the chital and hare, and that in both study areas domestic cattle, though readily available, comprised a very low percentage of the wild dog's diet.

COADAPTATION OF TIGER, LEOPARD, AND DHOLE

Johnsingh's (1979 & 1982) comparative study of the predatory behaviors of the wild dog, tiger, and leopard shows clearly how their hunting habits and prey selection help reduce competition between them, and, at the same time, maximize the utilization of available resources. Many factors contribute to the behavioral and ecological isolation of these three species. For example, Johnsingh found that wild dogs kill more chital fawns and aged chital while the panther kills more does and, also, more randomly. Tigers tended to kill prey in the 100–200 kilogram weight range, leopards, in the 50 kilogram or less range, and wild dogs, predominantly in the 25 kilogram or less and 25–50 kilogram range. Diversity of prey species and of age classes within species in a given biome enable a greater diversity of predatory species to share the same habitat.

Coadaptive factors other than prey diversity and prey selection can also be identified. These factors include the nocturnal hunting activity of leopard and tiger which contrasts with the diurnal or crepuscular activity of the dhole. These big cats need cover, hunt by stalking their prey, and are solitary hunters, while dholes course after their prey and hunt together. According to Johnsingh, other factors help isolate these species in time and space—notably, the leopard's arboreal proclivity, and the tiger and leopard's territoriality and habit of caching food.

METHOD OF KILLING AND DISSECTING PREY

We found six fresh kills (four chital and two domestic cattle) before vultures and other scavengers arrived, and noticed they were in varying degrees of mutilation. The various signs the dogs left enabled us to piece together how prey is brought down and dissected.

One of the dholes seizes the prey by the nose,* and the hold, like a twitch on a horse, must greatly inhibit the prey's movements. Other dogs attack the hind end, biting the thighs, buttocks, and flanks. One dog may grab the tail and, with the other animal on the nose, one or two dogs on the ears, and the rest of the pack pulling at the prey's flanks and hindquarters, a tug-of-war ensues. The prey is pulled down, and, if it is a fawn or calf, it is literally torn apart. There is no killing bite, as with the big cats, to swiftly dispatch the prey. The wild dogs, although they have powerful jaws, have relatively short canine teeth (see Figure 5.8) which cannot be used to sever the spinal cord of large prey. These shorter canine teeth, aided by the outer upper incisors that are curved and have evolved into a pair of secondary canines, enable the dogs to get a secure hold on the prey. The sight of several dogs securely attached to the prey and pulling, twisting, and tearing, has shocked many a hunter. They are indeed messy, having no clean and efficient way of killing their prey.

In contrast to other canids, the whistling dog has extremely well-developed masseter or jaw muscles, which gives its angular head an almost

*Similar behavior is seen in the African hunting dog (James Malcolm: personal communication 1981).

Figure 5.7 A large chital stag killed by four wild dogs; note torn nose from nose hold during kill. Village pi-dogs (pariahs) would soon clean up the remains.

hyenalike appearance. On palpation and dissection postmortem, this impressive difference in the muscles of mastication is apparent; as in other canids, the temporal muscles are also well developed. Such a specialized development of the masseter muscles enhances the wild dog's ability to hold

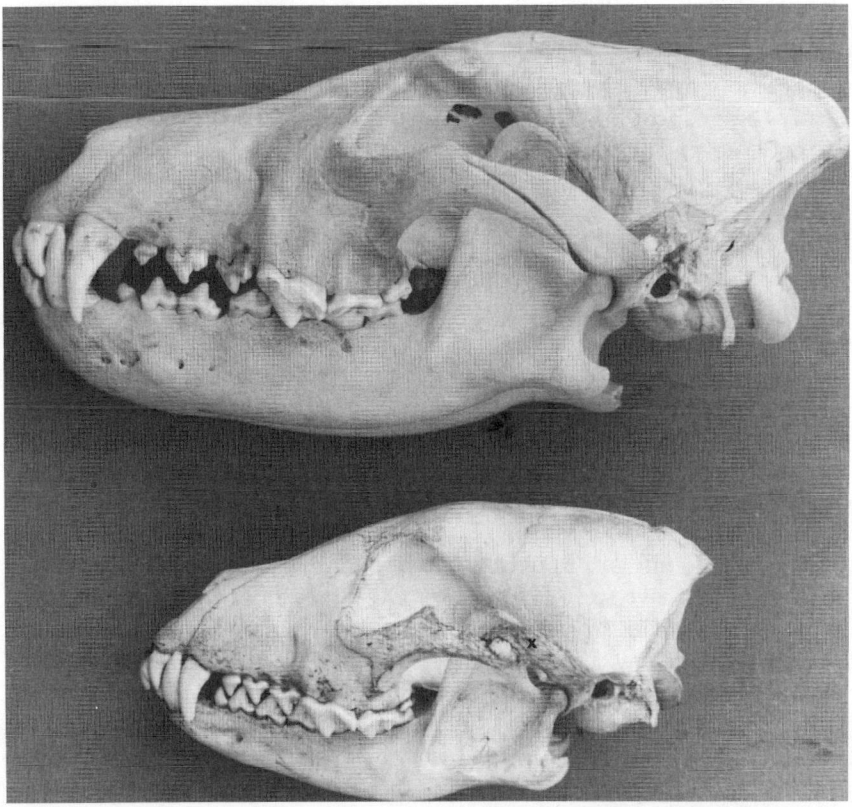

Figure 5.8 Comparison of wolf (above) and whistling dog (below) skulls; note more "catlike," shorter muzzle in latter species, and also healed fracture on the zygomatic arch (X), probably a hunting accident.

its prey securely and to crush the bones of the carcass, often ingested in considerable quantity.

Even so, the whistling dog does not have a clean killing bite and will often tear the flesh from its prey while the latter is still alive. Horrified by this mode of dealing with its prey, one Indian resident, active in humane animal care in the district, regards the wild dogs as "simply treacherous and vicious—it's the only animal that I will ever shoot in the jungle." Clearly, even educated people, especially those concerned with animal welfare, have to be taught to see the cold realities of nature without anthropomorphic judgment and to empathize with both the deer and the wild dog without moralizing and agonizing over the wild dogs' actions.

When the prey's flanks are torn open, the animal is eviscerated (disemboweled) as it struggles or is dragged along the ground by its head. Liver, kidneys, heart, and lungs may be eaten, and some sections of the intestines, but the stomach (abomasum) and rumen are torn out and left untouched.

Similarly, in a wild pig kill that was found, the stomach was untouched by the dogs.

Dholes may eat portions of the prey that are torn off as it struggles, hence the frequent observation that wild dogs will even eat their prey while it is still alive. The eyelids and eyes may also be eaten. Those people who find a fresh wild dog kill or see them making a kill have proposed that the dogs bite out the eyes of the deer and blind them first. This action is an unlikely probability, considering the difficulty a wild dog would have in seizing the eyeball, retracted deep into the orbit as a defensive reflex by powerful retractor oculi muscles. Rather, the dogs remove the eyes when the prey is dead or immobilized in shock prior to death.

Another lay observation supposes that whistling dogs whip their tails (soaked in urine!) into the eyes of their prey to blind them. This myth can easily be refuted! Wild dogs perform a body slam with the shoulder or hips to knock over the prey, and, undoubtedly, this action has been misinterpreted by some observers.

In the absence of a killing bite, what physical stimulus kills the prey? Young fawns suffering from multiple bite injuries have been rescued from wild dogs and have recovered; others have "played dead" (the tonic immobility reaction) and, suffering from less extensive injuries, were able to run off when the dogs were disturbed.* Normally, the prey probably goes into a state of shock, and death ensues rapidly after evisceration, no doubt, the major physical stimulus killing the prey. In the absence of evisceration, the onset of shock may be prolonged and the wild dogs would have to fight with the prey longer, possibly risking injuries to themselves during the attack. What more efficient method, then, in the absence of more effective weapons, than to disembowel the prey? An understanding of these facts will hopefully give the hunter and naturalist a clearer understanding of why the chennai seem to be such bloody killers.

DYNAMICS OF HUNTING ACTIVITIES

As Mech (1970) has shown, the North American wolf kills a very different portion of the deer and moose population than does the human hunter, who is either seeking a trophy animal or is limited in what he may shoot by local hunting regulations. The same is true in the Nilgiris: whereas a hunter may not kill fawns, does, or young stags and may overlook aged males whose antlers are "growing back," the wild dogs may kill opportunely, on a random or chance basis. Since herds in the springtime consist mainly of fawns and breeding adults, these are most often killed. Subadult deer, one to two years old, are rarely taken. Seemingly, the predation pattern of the wild dog is ecologically more adaptive than the human pattern of hunting

*E. R. C. Davidar 1974: personal communication.

only trophy stags. In the absence of wild dogs, human hunting would have to be changed, intensified, and shifted to include the fawn "surplus" (a surplus produced in "anticipation" of predation is the rule in all prey species) and the breeding adult does. Without such a change, the surviving surplus of fawns and adults would rapidly result in overpopulation, overgrazing, and, ultimately, enormous deer mortality from starvation and stress disease (aggravated by competition with indigenous domestic cattle; see Chapter 7). The population cycles of prey and predator that normally control the predator population and prevent overkill will be discussed later (Chapter 8).

The probability of seeing an actual whistling dog hunt was extremely low. A chance encounter during random searches through a known hunting area was all we could hope for. Our presence in the hunting area could have adversely affected the wild dogs' hunt by disturbing them and their prey, and field workers should be mindful of such an interfering variable. Some field workers in America are able to follow wolves by airplane and claim that neither wolves nor their prey (caribou, deer, or moose) are adversely affected because they soon become habituated to the plane. Even if we had had a plane and a radio collar on each wild dog, the vegetation, even in the relatively dry season, would have been too dense to permit aerial observation of hunting activities.

We were extremely fortunate, however, in seeing three hunts. In the first incident, three dogs were seen traveling in a crisscross pattern up river, working alternately along the river banks and crossing from one side to the other by leaping over the rocks. Suddenly, the lead dog froze and oriented toward something in the vegetation on the river's far side. Its reaction was so fast that the cue given by the prey was probably visual or auditory; a wild dog catching a scent usually persists in sniffing and searching until it can see or hear what its acute senses have detected.

The lead dog immediately charged into the labyrinth of bushes, but the two other dogs did not follow in its tracks. They fanned out in an arc behind, entering the bushes on the far bank and ready to cut off the prey if it bolted downstream. The time was 7:55 A.M. and the kill was made 20 yards from the river bank in almost impenetrable scrub, marked by the excited yipping, yakking, cackling sounds of the dogs and the scream of a fawn. By 8:30 A.M. several crows and vultures had gathered in the trees, but there was little left of the four-month-old fawn except some shredded skin, part of the skull, the rumen, the pelvic girdle, and both femurs of the hind legs that had been ripped away.

The second hunt was unsuccessful. Some 300 yards from their denning area in the hills above the valley, four dogs and a cub about ten weeks old were seen at 6:25 A.M. chasing a small herd of gaur, or Indian bison. They made the same excited sounds heard during the first hunt. Two cows kept the wild dogs away by simply standing their ground; between them were two calves. The dogs gave up after five minutes, returning downhill

Figure 5.9 Wild dog in a stalking run coordinates with hunting companions to drive sambar into river: three does, tails raised in alarm, stand at bay protectively around a fawn. (Photo courtesy of J. A. Cohen)

toward the dens in the direction of a series of shrill whistles. This whistling was probably the lead dog calling them back; they made soft whines in response to the whistling as they headed toward the sound.

That the wild dogs frequently kill their prey in the river where the water is often deep instead of killing in the open may indicate there is some advantage to driving prey into the water. In a third hunt we observed, a sambar fawn was driven to a rocky river crossing and then forced into the water until it was in a deep section submerged to its neck. The dogs then attacked. Other observers have even seen one or more dogs climbing onto the prey's back and biting at it as it floundered in deep water. The advantage would be in the dogs' favor, since the prey would soon become exhausted from struggling in deep water and would not be able to effectively strike out with its front feet in defense. Thus protected from the sambar's or chital's sharp front hooves, the deep water strategy may be more than a fortunate coincidence. Since several hunts occurred in the same regions along the river where there were deep pools, the wild dogs may deliberately herd their prey to these favored locations.

In one deep water river kill, eight dogs drove a screaming sambar fawn into the river at 7:50 A.M. Langurs in the surrounding trees were giving alarm calls and, within two minutes, crows were circling over the scene. While dogs held strategic positions along the river bank, presumably to prevent the fawn's escape and to keep it in deep water, three dogs attacked the fawn's head, one each behind an ear and another on its muzzle and above the right eye. The fawn was completely submerged for a few moments after this attack and two of the dogs departed, taking up positions along the bank. The fawn was attacked again by the third remaining dog, the largest in the pack. Attacks were directed again at the fawn's face, the rest of its body being submerged. The struggling fawn went underwater again, and this third dog then left it in the river. After a few minutes the dazed fawn began to give weak alarm calls and make its way toward the shallows. The third dog immediately threatened it from the bank and, facing up to it, uttered a series of cackling yelps, whereupon the fawn froze and remained in the shallows. Another dog dashed to attack, but was actually driven off by the large dog. The latter then drove the fawn into deep water and attacked its face again, and the fawn collapsed completely and went under. The large dog proceeded to drag the fawn toward one bank and was assisted in the water by the two dogs involved in the initial attack. Having reached the river bank, these two dogs left and three others helped the large dog pull

Figure 5.10 (1) Pack of whistling dogs hunting. (2) Lead dog keeps sambar fawn in deep water and (3) attacks, while two dogs in water and two on river bank stand by. (4) Two dogs in water help the leader drag fawn onto the bank and the rest of the hunting pack (three dogs) line up along the bank (6:30 A.M. kill). (Photo courtesy of B. Barnett)

the prey onto the bank. The distribution of three dogs along the banks, oriented away from the kill, gave the impression that they were acting as lookouts, possibly for the return of the fawn's mother.* One or more large sambar does can effectively drive off a pack of wild dogs. "Lookout" dogs might, then, serve as decoys and prevent a doe from intervening while her fawn is being killed. This particular kill took approximately ten minutes,

---

*Keller (1973) similarly observed whistling dogs acting as lookouts while the remainder of the pack were eating their kill.

although the fawn may have been chased some distance and could have already been exhausted by the time it entered the river.

Johnsingh (1979) observed a similar prey-capture and kill sequence, which he described as follows:

> Bent Ear (the presumed leader of the pack) caught hold of the rump of the fawn and tried to pull it towards the shore. But the deer stood firm. Then Bent Ear jumped on the back of the chital but slipped. However, he caught the muzzle of the fawn, swam and then walked, dragging the fawn all along to the shore. During this the fawn screamed six times. As the pair was about to reach the shore dholes, 4 other adults, stood alert to join the melee.
>
> When the chital was dragged to ankle deep water all those that stood ready sprang on it and there was a scream. The chital was dragged for 3–4 metres away from water and fully eviscerated on the way.

One is again impressed by the absence of a killing bite and the fact that the prey was not eviscerated until it was in shallow water, since in deep water disembowelment would have been impossible. Bites were confined to the lips, nose, periorbital region, and behind the ears. Hypothetically, such repeated trauma to the sensitive areas of the head brings on a rapid state of shock as with evisceration, leading to a vasovagal attack and "fainting" (the facial and trigeminal nerves are extremely sensitive). Abrupt recovery from such a fainting reaction may explain why people have occasionally seen the prey suddenly struggling while the dogs are eating it—not all prey dies from shock before being eaten.

In his thesis, Johnsingh includes valuable additional material on the wild dogs' hunting behavior. Most prey weighing 50 kilograms or less were killed within two minutes after first physical contact. The maximum time to kill a large stag was fifteen minutes. Of forty fresh kills, twenty-seven showed signs of having been held by the nose. The wild dogs held the nose in all cases when killing sambar stags and in 94 percent of chital stags in hard antler, compared to 25 percent of chital doe kills. No throat bites were seen. The wild dogs appeared to enjoy the prey's eyeballs, twenty-one of forty fresh kills having the eyes removed. The dogs do not lie around the kill eating, but usually tear off a piece and run to cover to eat alone. This behavior may well be to avoid possible human encounters as much as to avoid intrapack conflicts while eating.

While most kills occurred quickly, an unusually drawn out one was vividly described by Johnsingh in his thesis (1979):

> *Hunt 9. 24.3 1977*
>
> In the evening I was seated on the top of the shooting hide. At 1730, 100 metres west to the shooting hide, 7 dholes were seen attacking a chital stag with 90 cm long hard antlers. They had to be observed only through the tree trunks. By biting and hanging on the rump 2 or 3 dholes had made the stag almost immobile and 2 or 3 dholes were worrying the flanks. The stag tried to keep away the dholes biting on the flanks by swinging its antlers. This it did 5 times but the antlers never touched the dholes.
>
> At 1733 one dhole caught the snout of the stag and started pulling the deer forward while those at the rump continued to bite and pull backwards. Now the stag was completely helpless and by 1735 it was brought down. During this struggle the deer never screamed aloud and its sounds made thrice were feeble. The dholes were absolutely silent during the attack.
>
> The stag, keeping its head erect, lay down on its side. The dholes were eating the stag alive. Twice it attempted to rise but the dholes eating at the rump and the flanks prevented it from getting up.
>
> At 1737 I got down from the hide and crept a little forward and for the first time my presence was felt and thereafter the dholes were a bit disturbed. While 2 or 3 ate the live stag others looked all around for the source of disturbance. One dhole climbed a 1.5 metre high termite mound and kept watch. At 1745 the stag's head dropped sideways on the ground and he appeared to

> be quite dead. It had taken nearly 15 minutes for the stag to die. Most probably my presence could have disturbed and delayed the killing. When the carcass was checked later it was found that the left pinna had been bitten off from the base and both the eyes were uninjured. Viscera lay 5–6 metres away. Both the testicles without skin lay close by. [Johnsingh regards such emasculation coincidental to attacking the rump and flanks.]

Other chance encounters of wild dog hunts by reliable observers are few. Prater (1965) notes that when the pack is close to the prey, the dogs may utter excited whimpering sounds and break into a clamorous yapping when the prey is close and in view. He also notes that families often unite to form larger hunting bands in the rainy season. Large hunting parties of twenty or more dogs may be seen at this time of the year. But during our period of study, the two packs of approximately twenty and twenty-five dogs we followed were only rarely together and not when hunting. The hill pack, early in the morning, broke into smaller hunting packs of three or more dogs and reassembled in the evenings at the denning area. The second pack similarly broke up to hunt in the early morning but reunited after the hunt to spend the rest of the day by the river near the denning area.

Packs of whistling dogs have been known to kill tiger and sloth bear. Our study confirmed that they will kill adult wild pig and possibly leopard (or panther) since the chewed skull and other bones of the latter predator were found in the nursery play area at the den site of another pack we visited. Killing the large cats must be extremely rare, for a leopard will usually seek refuge in a tree. Johnsingh (1982) reported the death of one cub attributable to a leopard. While he saw no evidence of leopard or tiger competing with dholes over kills, the dogs did scavenge on leopard kills and occasionally on tiger kills. Davidar (personal communication 1974) has seen several wild dogs relaxing along a river bank while a tiger ate its kill, and the big cat was quite unperturbed by the dogs' presence. Similarly, he has seen chital and sambar come to drink with wild dogs in full view close by. As with deer, caribou, and wolf, if the predator is obviously relaxed, the prey will virtually ignore it and continue to browse or drink. You wonder why the wolf and wild dog have not evolved some adaptive strategy of appearing to be relaxed and nonchalant in the presence of prey when hunting (Lois Crisler in *Arctic Wild* [1958] notes how effectively a relaxing stretch and yawn by a human will put a fearful wolf at ease). More likely, the potential prey recognizes that the hunters are relaxed because of the time of day.

Pythian-Adams (1949) shows that wild dogs will kill langur monkeys and, from fecal analysis, they will seemingly kill anything: they are opportunists. But they have an obvious aversion to the most easily obtainable prey—domestic cattle and their calves.

A sambar, which may be twenty times heavier than a wild dog, will be hunted if it can be singled out from the herd. If two or more sambar at bay

Figure 5.11 Contrast in size of adult sambar stag and whistling dog (redrawn from Prater 1965).

stand back to back, they can keep the dogs off effectively and they will collectively defend one of their fawns. The wild dog strategy is to try to separate the fawn from the herd where it may then be easily taken. Both chital and sambar kills are often found near water, leading to the common deduction that deer at bay will run to water in an attempt to escape. Many kills, however, are made when deer come to water or to habitual crossing points along the river used while traveling from one browsing area to another. Another reason for so many kills near a river is that the prey takes the swiftest route away from the dogs—downhill where, coincidentally, the river lies.

On the gently sloping bank of a lake, adjacent to a dam in the west corner of the study area, many chital kills were found. In one clearing, we discovered the remains of five deer killed over a five- to six-month time period, judging from the bleaching and decalcification of the bones. In two other clearings, six kills were found, all within 5–15 yards of the water. Such consistency in the place where kills are made points to some habit of the prey such as that alluded to earlier. Possibly, the wild dogs use the predictable behavior of prey in their hunting strategy. Thus, some dogs could wait in ambush near the water while others drive the deer toward them. We were not able to confirm this hypothesis, although wolves will use such tactics in driving caribou and Dall sheep toward other wolves hiding in ambush downhill or further down the valley.

On a long chase, one wild dog and then another may reportedly take over in close pursuit of the prey, while the rest of the pack keeps up a steadier pace behind. This "running in relays" may be a misinterpretation of canid hunting behavior, and has been clarified somewhat by Hugo and Jane Van Lawick's observations of the Cape hunting dog in their book *The Innocent Killers* (1970). As the prey zigzags in front of the pack, it comes closer to some dogs than to others; the closest dog takes up the chase until the prey again turns and closer dogs take over. The open plains of the Serengeti, where the Cape hunting dogs live, is very different from the jungle habitat of the whistling dog, although a few small clearings have the parklike appearance of this great African plain. Consequently, the hunting strategy of the whistling dogs would be adapted to their terrain. Running in relays would occur when the prey in front of the lead dog turns wide and dogs in the rear could intercept it by taking a shortcut via one of the many narrow game trails creating a labyrinth through the dense scrub.

In his thesis, Johnsingh (1979) writes that:

> Dholes were seen to adopt two strategies to hunt. One was to kill prey as they moved through the scrub and the other was to keep some pack members in the periphery of the scrub to intercept the fleeing prey flushed out by other members of the pack. Both these formations would be useful to locate hidden fawns and to flush out the resting animals. In total dholes were seen adopting the former strategy 39 times and the latter 14 times.

Most prey-chases were short, ending within 500 meters. If the food was not sufficient to satisfy the pack (estimated by Johnsingh to be 2 kilograms of meat per animal), a second kill would be made, often in the evening. As soon as the kill is eaten, whistling dogs move off quickly, either to a favorite resting place or to water. When there are cubs, the hunters may join them at a temporary nursery or rendevous site and feed them by regurgitation. When the cubs are still too young to travel and are confined to the den, members of the pack will return to feed them as well as the lactating mother or mothers. Guards or "baby-sitters," left by the hunting party and remaining with the cubs during the day, are also fed by regurgitation.

Rarely will dogs carry large pieces of prey back to the den, but this seems to vary from pack to pack. Of five denning areas visited, each belonging to different packs in the area, three of them had the bones and/or skin of prey near the den holes. This behavior may either be an idiosyncrasy of certain packs or purely a function of how close to the dens a kill is made; if it is close, some large pieces of the dismembered prey may be transported

Figure 5.12 "Graveyard" appearance of two areas littered with deer bones where wild dogs regularly make kills. Also, a wild dog dunging area where several scats (X) have been deposited, characteristically, at an intersection of two game trails.

back to the cubs. Johnsingh, however, contends that such remains around dens belonged to hyenas and that wild dogs do not carry the remains of kills to their den sites for their cubs.

Social behavior of wild dogs over a kill gives clear evidence of leadership and dominance relationships. In one kill observed, the leader of the kill was the first to eat, but within seconds was joined by four other dogs of a pack of eight; three dogs remained some 10–25 feet from the carcass, alert and oriented away from the other dogs, acting, as they did during the kill, as lookouts. Other dogs, after having fed, took the lookouts' places and the original sentinels then partook of the kill.

During a forty-five-minute feeding period, seven aggressive interactions were recorded. Five of these were threats, the more dominant dog making eye contact and, with an aggressive pucker, giving whine-growls toward a lower ranking dog. The latter dog avoided eye contact, crouched with ears flattened and lips pulled back into a submissive grin, and uttered yelp-whines. Two other agonistic interactions involved physical contact, the

Figure 5.13 Whistling dog "sentinel" keeping watch while other pack members eat their kill; it will soon be replaced by another and will return to the kill and eat.

Figure 5.14 Wild dogs at kill. (Photo courtesy of A. J. T. Johnsingh)

Figure 5.15 Wild dogs at kill. (Photo courtesy of A. J. T. Johnsingh)

dominant animal giving an inhibited bite toward the neck-scruff region; on both occasions, the other animal rolled over into lateral recumbency and, displaying a submissive grin, gave several yelp-whines. The aggressors in these two instances displayed a marked pilo-erection (hair standing on end) and the tail was held vertically erect (as in the wolf).

The feeding period was noisy, with vocalizations ranging from growls, coughing, and hacking threats and growl-whines to more submissive and distressing sounds of whining, yelping, yelp-whines, and growl-screams.

Tension eased after all had eaten, and two or more dogs often engaged in brief bouts of play-fighting and play-hunting. Some dogs may gape or gag after feeding, most probably a reaction to the prey's fur they have ingested (and which occurs in copious quantities in their feces). This interpretation is different from the common view that they are "belching" in order to relieve gas in their stomachs.

Chapter 6

FAMILY LIFE AND COMMUNICATION

Johnsingh's (1979) field research has added a few more pieces to the jigsaw puzzle of the whistling dog's social life. In the pack that he studied, up to eighteen dogs were seen together prior to the breeding season. A lone bitch and two pups joined the pack temporarily. He found that after November, around the denning season, the pack split, some seven to eleven members of the pack stayed in the core area around the den site while the rest of the pack emigrated. The economics of this split, in terms of not overpopulating the denning area and reducing the availability of prey needed to successfully raise a litter, are obvious; but the mechanisms that lead to this splitting of the pack and the factors which determine what animals stay at the den site, remain to be determined.

According to Johnsingh, the density of wild dogs in the Bandipur sanctuary, adjacent to the area covered in this field study, was estimated at 0.35–0.90 animals per square kilometer. He found that a pack of about thirteen dogs intensely used an area of approximately 20 square kilometers, covering a maximum area of 40 square kilometers and ranging daily from 1–8 kilometers. The adult sex ratio consistently favored males, with a ratio of 2:1. During the denning season, the hunting range was reduced to 11 square kilometers, while at other times it was over 15 square kilometers. The pack Johnsingh studied rarely split into smaller groups however. A large pack size may be advantageous in some locales in order to bring down larger prey and to take over the kills of tiger and leopard, which he observed on one and seven occasions, respectively: two or three dogs would be unlikely to intimidate these larger predators.

The breeding season of the whistling dog extends over a considerable period of time. There were three reports of observed matings in the first, second, and fourth weeks of November; one animal examined postmortem contained eight well-developed cubs, placing the time of conception around mid-October. With a nine-week gestation, the denning period would therefore extend any time from mid-December through February and into March, until late-born litters are weaned and able to travel with the pack.

Reggie Davidar, my naturalist friend whose cabin we used as a base camp in the Nilgiris, observed mating in a pack of wild dogs. The female assumed a crouched position ("like a lioness") with the male on top. At the moment of copulation, the female whined and six dogs ran over and submissively greeted the mating couple. Two of these, which Davidar thought to be subadults, stayed close by. There was no tail-to-tail tug-of-war as is usually seen in canids after the male dismounts with his penis still in the vagina. Instead, the pair laid down "flat on their sides, their bodies forming a semicircle facing each other" (Davidar, personal communication 1980). This resting lasted some seven minutes, during which time they nuzzled and "grinned" at each other (a distinctive facial expression, discussed later in this chapter). As soon as they broke apart, the male ran off yelping, attracting the rest of the pack. Three subadults and two adults surrounded the female. One adult sniffed her genitals and then moved off while the three younger animals showed great interest and investigated and licked the female all over as she laid down. Davidar concluded that, as in the wolf pack, copulation was an activity not necessarily segregated from the pack, but within the group by a selected pair. He saw no aggression, although the breeding male was not the largest member of the pack. In a captive pair in Duisburg Zoo, West Germany, colleague J. Cohen (personal communication 1979) observed a male marking its mate with urine, a behavior frequently seen in other canid species (see Figure 6.1).

FAMILY LIFE AND PUP DEVELOPMENT

There are few details in the literature on the behavior of wild dogs at the den, but our studies of different denning sites helped reduce this deficiency. At one active den site, three adults remained behind during the day while the rest of the pack, some seventeen dogs, went off to hunt. These latter animals split into smaller hunting packs, returning to the den site around 6:30 P.M.

Johnsingh was fortunate to be able to locate and study a litter of wild dogs and to chart the development of the pups (see Table 6.1).

Table 6.1 *Development of Pups*

31.12.76. Nearly 20 days— one pup sooty grown came out of the den - carried in by adult male.
5.2.77. Nearly 56 days old—pups actively chased and played with the mother.
13.2.77. 64 days—pups were alone near the den—black hair was seen growing on the tail; pups gave a growl-bark, then ran toward the den when disturbed by tourists riding an elephant.
15.2.77. 66 days—speed and agility while chasing and playing with a young male were remarkable.
18.2.77. 69 days—left the den.
20.2.77. 71 days—nearly 10 workers attempted to catch the pups—pups escaped.
3.3.77. 82 days—pups followed the pack to hunt—looked like full grown miniature dholes.
24.3.77. 103 days—pups were brought to the kills, 47 minutes after kill was made.
6.5.77. Observation of Keechanna my native field assistant—when the pack killed a Chital stag the pups came after "some time" in the company of 2 adults.
17.5.77. At 0725 I heard a Chital being killed—when the kill was found at 0732 the entire pack was with the pups—pups covered the kill—adults occasionally went and got some meat to eat it away from the pack.
21.6.77. Pups assisted Bent Ear to kill a big chital fawn.

SOURCE: Johnsingh 1979.

Their rate of development is remarkably fast compared to the wolf, closer to the coyote in many respects. Johnsingh found that the pups stay around the den site until they are seventy to eighty days old. Then they follow (or are led by) the adults to the kills. They help kill prey by the time they are six months old and by eight months will even attack sambar. Johnsingh (1982) reported that pups are suckled until they are at least fifty-eight days old, and do not use water holes near the den during this time. (Adults also fed the mother while she was confined to the denning area nursing her cubs.) After the pups left the den, they were frequently fed by adults regurgitating meat for them, which the pups would solicit by chasing and biting the corners of the adults' mouths. Unlike the wolf, the dholes do not use rendezvous sites to meet the cubs. However, one or more adults usually stays with the young and accompanies them as they move around the central den area.

Johnsingh has gathered further data on what goes on around an active wild dog den. A guard dog is sometimes present, especially in the evening. Interestingly, the dogs seem to avoid killing prey close to the den, an observation also made by researchers of the African wild dog and Australian dingo. Johnsingh hypothesizes that this behavior may reduce local dispersal of prey, thereby providing training opportunities for pups at a later age, supplying a buffer zone, and preserving a food reserve. Since local people frequently took the wild dog's prey, not killing prey near the den might help conceal the presence of an active den site.

Figure 6.1 Male marking mate with own urine, Duisburg Zoo. (Photo courtesy of J. Cohen) (see Cohen, J. A. [in press]. A note on the behavior of captive dholes [Canidae: cuon alpinus]. *J. Bombay Nat. Hist. Soc.*

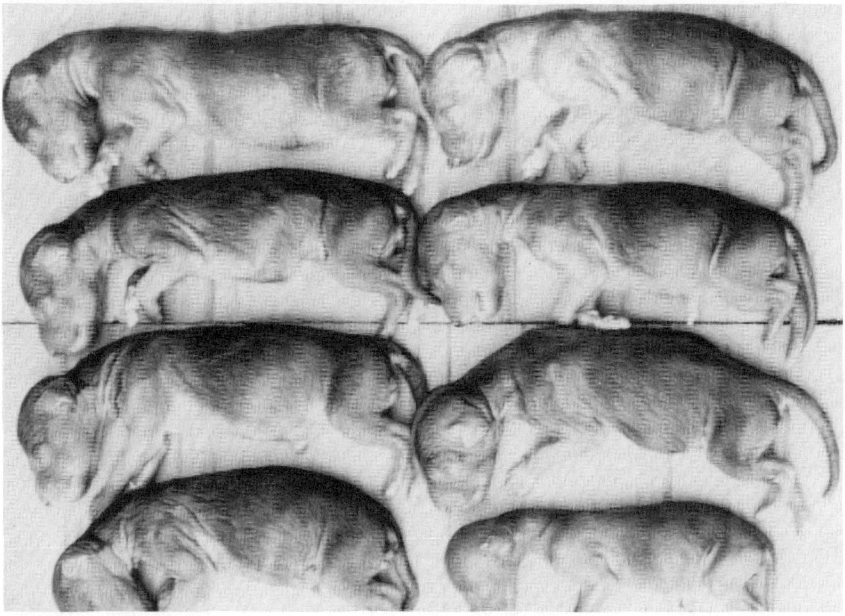

Figure 6.2 Attestation of high fecundity: eight cubs found on postmortem of three-year-old whistling dog.

Figure 6.3 Dholes ages approximately four weeks (above) and twelve weeks (below). (Photos: courtesy of E. R. C. Davidar)

PACKS AND CLANS

A better term for a wild dog group would be "clan" rather than "pack," since by general usage the term "pack" refers to a group of animals that always hunt together. The wild dogs of a "clan" break into small packs of three to five animals, especially in the spring dry season, this being an optimal number for taking chital fawns. Bigger packs may be seen at this time, but may be more common later in the year hunting juvenile and adult sambar and chital. From twenty-four sightings of wild dogs by us and graziers in the area between January and February 1974, we were able to estimate the average group size of hunting parties (see Figure 6.4). These few sightings indicate that whistling dog clans normally break into small hunting packs of about three animals at this time of year. The two sightings of approximately twenty dogs with six cubs and of twenty-four adults were, respectively, an assembly of animals after the morning hunt and a reunion at the denning area in the evening. These figures reflect the flexibility of the two clans in our study area, which numbered approximately twenty-six and twenty-four each. Very similar figures of hunting packs were also collected in the second study (see Figure 6.5).

Confirming the present findings of pack numbers, Saratchandra and Gadgil (1975) found that, in the Bandipur sanctuary adjacent to the Mudumalai study area, the most frequently observed "pack" size was three. They reported sightings as follows:

Pack size	*Number of sightings*
1	3
2	5
3	8
4	4
5	0
6	2
10	3
20	1
30	1

A total of twenty-seven sightings over a twelve-month period attests to the elusiveness of these hunters.

As I have discussed, the returning hunters of the pack feed cubs, lactating mothers, and den guards or baby-sitters by regurgitating meat. At a complex denning area or "city," the clan may have several litters of different ages to provide for. Davidar (personal communication 1975) has observed different dogs visiting a simple den to feed a lactating bitch which ran out and submissively greeted them in a food-soliciting manner very similar to the wolf. No particular male seemed to be attached to this female, and, probably providing food for cubs and mothers is a communal activity. By the time cubs are eight to ten weeks of age, they may be moved to another den site or to temporary nurseries of hideouts in a different part of the hunting range.

Unfortunately, we were able to learn nothing about the relationships between different clans of wild dogs. Questions such as whether they have territorial boundaries, if there is aggression or mutual avoidance between

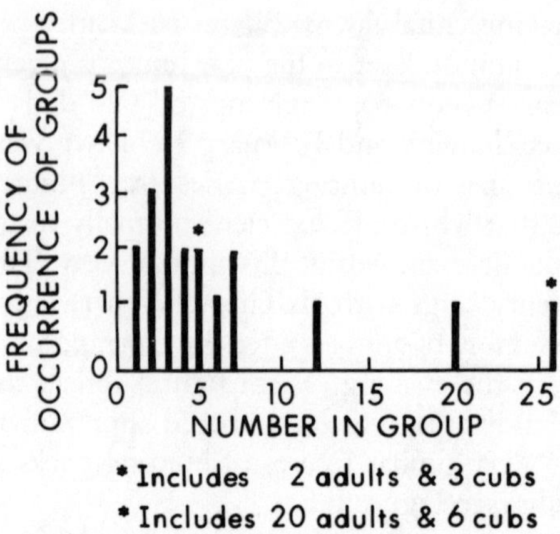

Figure 6.4 The most frequent size of whistling dog hunting parties seen during the dry season was three; groups of twenty and twenty-five were assembled after hunting. (From direct observations and graziers' reports.)

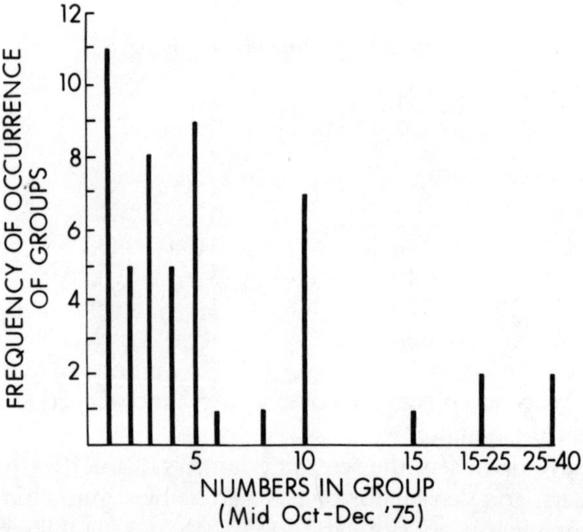

Figure 6.5 Whistling dogs were most often seen singly or in small hunting groups of three-five during the end of the monsoons and onset of their denning season.

clans, and whether or not a clan or certain age class of the clan will unite with another clan at certain times of the year, await further research. Although there is still an abundance of food in this Nilgiri valley, staying within a particular hunting range and avoiding contact with neighboring clans could be an important social behavior essential for ecological equilibrium. An excessively large clan might overkill in its own hunting range and, if adjacent clans did not keep this clan out of their own territories, the eco-

Figure 6.6 A hunting pack on the move. (Photo courtesy of A. J. T. Johnsingh)

logical consequences could be disastrous for the whole wild dog population. Confined by territorial limitations, an excessively large clan would regulate its numbers by starvation, higher pup mortalities, and lower reproductive success, or some members might emigrate. But, at this stage, such notions are sheer speculation.

In the two assembled clans that were seen, there was one large animal in each that was the most alert and appeared to cue the others by moving off in a particular direction or by attracting attention with their behavior (investigating, sniffing, looking, and growling at human intrusion and rallying others with barks). Burton (1941) reported that in a litter of hand-raised cubs, one was clearly the most dominant, to which the others responded as though it were a leader.

The question then arises, if there is one clan leader, are there also others of high rank or of a certain age and experience who lead the small hunting packs? It is quite likely that there is less of a dominance hierarchy or rank order in the clan than we see in the wolf pack, and that the whistling dogs are therefore comparable to the Cape hunting dogs. In an environment abundant in food, there may be no need for the more rigid social structure and territoriality manifested in the wolf pack. Field work with primates supports this interrelationship between ecology (abundance and distribution of food) and social behavior. Gorillas, for example, living in a tropical abundance of food have a relatively relaxed and open social structure with free-flowing relationships and exchanges with other groups.

Figure 6.7 Members of a small, wild dog hunting pack. (Photos courtesy of E. R. C. Davidar)

In his independent field study, continued after the present study was terminated, Johnsingh collected additional data on the social behavior of the dhole. He saw no dominance displays (like those described in the wolf pack) by whom he identified as the pack leader, although other dholes showed frequent active submission toward this individual. He found that pups were "quarrelsome" and subadults frequently fought, but adults were amicable, especially toward the pups. The latter were allowed to eat first at kills. All adults regurgitated food for the mother and also for the pups. He found no evidence of the dogs carrying bones or pieces of prey to the den. He saw pups as old as six months of age soliciting food and being fed by regurgitation.

Even local people who dislike whistling dogs, considering them needless killers, are impressed by the lack of competition at a kill. "Everyone gets a share, they really do take care of each other" was one comment. In contrast, in the wolf pack the more dominant members eat first—a highly adaptive social ritual in an environment where food may be seasonally scarce: the fittest survive and reproduce. Seasonal scarcity is not usually a serious problem for the Cape hunting dog or whistling dog, so the need for a rigid dominance order while feeding is unnecessary. The Van Lawicks (1970) noted that in the Cape hunting dog, cubs are allowed to eat first. This canid, like the whistling dog, may have more than one bitch in the clan raising a litter of cubs. Such a situation is extremely rare in the wolf pack; usually only the dominant female mates, since there would normally be insufficient food for more than one litter per year.

With noncompetitive food sharing at the kill, with the common occurrence of several bitches bearing cubs in the clan, and with communal care of the young, how then is the wild dog population regulated?* In the wolf, regulation seems to occur prenatally (or rather, prior to copulation) in that the most dominant female prevents others from breeding, i.e. a social regulation of population (Fox 1971). According to the Van Lawicks' documentary film, *The Innocent Killers,* a dominant female Cape hunting dog may kill the cubs of a lower-ranking female if food is scarce and if the cubs are much younger than the dominant's own cubs and unable to travel with the pack. Interestingly, in the Serengeti lion sexual behavior is extremely frequent—an unusual phenomenon in cats that may have a secondary social function in keeping the pride together. As a consequence, several lionesses may bear cubs and once again the problem of population regulation arises. According to Schaller (1972), when food is scarce many cubs starve to death since the first to eat at the kill are the most dominant members of the pride.

It is not known precisely how, and at what stage of cub development, population regulation takes place in the whistling dog when food is scarce or when the population is very high. Further field work will show whether the animals practice infanticide like the Cape hunting dogs may, or if a dominance-rank competition develops as in the wolf when feeding. Johnsingh (1982) notes that

> The size of the kill may influence the degree of tolerance shown to the pups by adults at kills. When the kill was as small as a chital female the pups monopolized the kill and only occasionally the adults took pieces of meat (n = 3). When a yearling male sambar, 2–5 times larger than a chital doe was killed, the eight six-month-old pups and eight adults fed together.

---

*As Johnsingh (1982) implies, the pack may split during the breeding season and only the remaining dominant female breeds. When pack tension is high and splitting occurs, the availability of prey may be low thus ruling out the possibility of several bitches with cubs sharing the same den area.

Figure 6.8 Two elusive and wary wild dogs on early morning hunt (upper row). Two dogs act as guards near a den of cubs while the main pack is out hunting (lower row).

A precopulatory control, manifested by a dominant female emerging through competition for food when scarce during the breeding season, is unlikely. The possibility, however, must be entertained, as Johnsingh's study implies.

Most likely, each clan is a finite number, and cubs normally fill the

ranks of those adults dying from disease or physical injuries inflicted while hunting. The turnover rate will be extremely high in some years and the average age of clans considerably lower when distemper epidemics contracted from pi-dogs erupt. This loss would subsequently be compensated for by a greater number of cubs surviving to maturity and occupying the niches vacated by the earlier generation.

Without further data we can only speculate about the behavioral significance of "clan" formation in the whistling jungle dog. The underlying

Figure 6.9 A whistling dog, acting as den guard, reclines and stretches at sunrise while the pack is away hunting (6:15 A.M.).

factors which may facilitate this phenomenon have been outlined, but the question remains as to the adaptive significance of such behavior. One noted behavioral ecologist, Wynne-Edwards (1962), has described the often brief social aggregations in many diverse species as possibly related to a population regulation feedback process, a phenomenon he terms an *epideictic* display. Such displays may trigger reproductive cycling or inhibition thereof, or lead to new groups forming and dispersing. While this phenomenon is common, its purposes and consequences are not fully understood. In the wild dogs, social interactions during clan assembly may influence population dispersal and reproduction. We must, however, be careful in ascribing specific purpose or function to a particular behavior. "Clanning" in the wild dog may be a purely pleasurable social activity rewarding in itself with no reproductive or other ecologically related purpose or function.

While doing a study of feral and free-roaming urban dogs in the United States, I observed a comparable (but extremely short-lasting) phenomenon to clanning. Several dogs in the neighborhood came together for several minutes and engaged in mutual sniffing and friendly greeting, then they dispersed to forage for food separately. These early morning "group greeting rituals" may have been coincidental, the separate dogs being in the same place at the same time. Similarly, separate family packs of wild dogs sharing the same hunting range may chance to meet together at one of their many favored hunting spots along a river or in an open glade. Such meetings would help maintain social bonds between neighboring family packs and be the foundation for clan formation. If these speculations have any

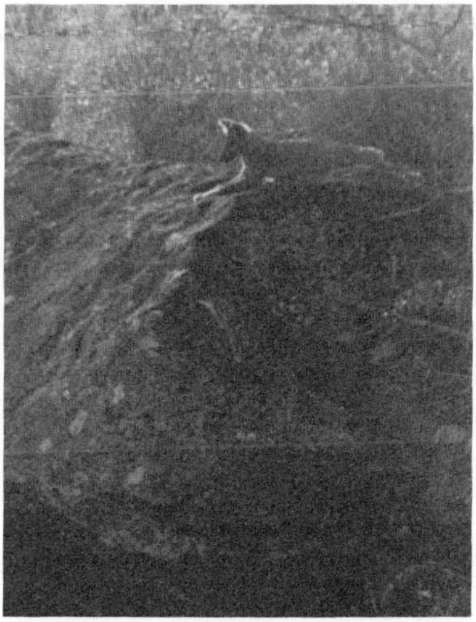

substance at all, the nature of the relationships between neighboring wild dog clans remains very intriguing.

This major question concerning the clanning of the chennai remains to be answered and the answer will not be forthcoming without intensive study of animals trapped and released wearing radiocollar transmitters. We may never have the answer before these extraordinary and little-understood (or appreciated) creatures are extinct. The myths of the vicious and sadistic ways of the wild dog still persist: an article about our work in one of the national newspapers of India during our study there contained quotations from a forest official. He recounted the animals' cruel ways of tearing out the eyes of their prey to blind them before they are eaten and told how plentiful the dogs are, ravaging the countryside in huge packs.

Huge packs: herein lies the mystery that remains to be solved. Most sightings of whistling dogs are of small groups, usually three to five and occasionally eight to twelve, and these are hunting packs. More rarely, larger aggregations of twenty or more are seen together, resting, playing, and interacting socially, but rarely hunting together in such large numbers. Davidar (1974) describes his observations at the dens of the dhole, and herein lies a clue. Around one active den he observed twelve animals. Two were yearling females which behaved submissively toward two adult lactating females; a third pregnant female was also in this group, sharing the communal den or nursery. The second denning area was occupied by one lactating female which solicited food, regurgitated by a large male and by other dogs. In this latter group, eight dogs, four of which were females, were seen.

Each hunting group of wild dogs may well be a family unit of parents and their surviving offspring, the latter, as in the wolf, assisting the parents in rearing another litter the next breeding season. With sexual maturity, some offspring may then leave the family group and join compatible offspring from neighboring families. This "exogamy" or flow between family units would be enhanced by a lack of territoriality afforded by a year-round abundance of food. Wolf packs, in contrast, tend to be more territorial and disputes over a territorial hunting range have been observed by Mech (1970) when prey was scarce; he observed one wolf pack drive out and occupy the adjoining territory of another pack. Such tension between family-unit packs and the need for a territorial type of social organization would not be necessary for the wild dog because of the abundance of prey. Relationships with neighboring family units would consequently be more relaxed, increasing the probabilities of exogamous matings (endogamous matings are usual in the wolf). The "openness" of family units may therefore facilitate clan formation in the wild dog, where different families come together at certain times. Mixed hunting groups may be temporarily formed, as well as communal nursery arrangements at favorable den sites. The factors leading to clan aggregations remain to be identified; large aggregations of dogs are seen at all times of the year, with a possible trend toward increased gregariousness during the breeding and denning seasons.

My research with captive wolf packs at the Naval Arctic Research Laboratory, Point Barrow, Alaska (Fox et. al. 1974), gives one clue possibly relevant to understanding the social dynamics of the chennai. While working with two packs of wolves aged one and two years, respectively, I saw that members of the younger group, when introduced into the older group's enclosure, were attacked (but not injured) and not accepted. When placed into the yearlings' enclosure, the older wolves were again dominant; but, as soon as an even older wolf was introduced, they submitted and showed signs of gaining acceptance into the younger pack. This acceptance was attributed to the catalytic effect of the nine-year-old wolf, with whom both packs had been raised for one year and who they regarded as their pack leader. I therefore theorized that, under natural conditions, an older animal may facilitate social interactions and group formation between separate packs that have never met before if such an individual was dominant or a leader figure. The joining of wild dog pack families could be similarly facilitated by one or more dominant individuals known to some of the other higher ranking dogs but not, of course, to their cubs. The cubs would be readily socialized and assimilated into the local clan via kinship bonds between and across different generations. This hypothesis gains some support since in both large clan assemblies we observed there was one individual of outstanding size and of dominant status who had a clear leader role.

That large clans and smaller hunting packs and family units are seen in the same area indicates that the wild dog is a highly flexible animal socially, but the hows and whys of this pattern of social behavior remain to be elucidated. Although the wild dogs leave plenty of tracks and scats, studying their social behavior and observing close social interactions remained unfulfilled expectations that were a constant source of frustration in our study. The accounts of chance encounters by graziers and other people in the jungle are not reliable, nor are the anecdotes of hunters, guides (shikaris), and a few naturalists. But any consistency in different reports gains validity as more reports and anecdotes are collected and scrutinized. Much of the material pertaining to the whistling dog in this chapter comes from such sources, which, in addition to our own field data, will have to suffice until more direct and continuous observations can be made.

Trapping animals and releasing them with radiotelemetry devices, a very costly method, is perhaps the only feasible way to closely study wild dogs' behavior (as Mech has shown in his studies of the wolf in Minnesota [1970]). Before this procedure is performed on whistling dogs, however, the animals may be extinct and their habitat no longer conducive to their continued existence or reintroduction from captive breeding stock. No breeding stock exists in captivity to date, save for a few in a Moscow zoo and at Duisburg Zoo in West Germany whose viability is questionable. Although the ethics of maintaining threatened and endangered species in captivity is debatable, this tactic may be the only viable alternative for the whistling dog. No scientific study in captivity can give all the information relevant to behavior and ecology in the wild or to improved wildlife management; at best, captive studies merely support field work by gaining a few insights and in clarifying certain issues unanswerable in the field. Clearly much more work is needed on the social organization and dynamics within and between wild dog clans.

COMMUNICATION

Some of the signals wild dogs use to communicate with each other have been studied and their "body language" provides further insights into their highly evolved social lives.

The most remarkable of the wild dog's repertoire is its whistle. How it produces this sound remains an enigma. The closest known sound from the canid world is the "coo-coo" of the red fox.

We witnessed this whistle on the only time we succeeded in making direct contact with the dogs by following fresh tracks. We suddenly found ourselves surrounded by three or four dogs closing in on a sambar hiding near where we also were hiding. The sounds the wild dogs made gave us empathy for the prey. A shattering "KaKaKaKAA" scream was emitted by one dog, presumably as it found and faced the sambar. Then the dogs began to whistle to each other in long trains of toots. These sounds perhaps helped the dogs coordinate their movements and gave each other information as to their spatial distribution in the dense scrub, since they could not see each other. Temporal information was also given; we could visualize a dog moving only a few yards away by its train of whistles, and yet it was quite invisible.

Previously, we thought the chennai whistled to locate each other and assemble as a pack, but now, being virtually in the role of the prey, we experienced this remarkable adaptation to hunting in dense cover where visibility is restricted to a few feet (and at that time there was virtually no breeze to give a good scent direction to the prey). The jungle whistling dog has evolved a repetitive, whistlelike, high pitched howl to send spatial and temporal information to pack mates who can respond appropriately and so coordinate their hunting without visual contact with each other or even with the prey. Although a favorable wind, clear track, or spoor, with one dog giving a scream call when facing the prey, help orient the dogs to their quarry, it is vital for them to encircle the prey, preventing its escape, and to coordinate the attack so they all strike in unison. This hunting strategy would be especially important when hunting such a large and powerful animal as a sambar. Hence, their special means of communication help make them successful hunters in dense underbrush, and also unique among the *canidae*.*

While Krishnan (1972) reported hearing the dholes whistling while pursuing prey in dense brush, Johnsingh (1979) contends the wild dogs' whistle is a contact call "used to reassemble after the pack members got dispersed after an unsuccessful hunt or by a vehicle." Both interpretations may be correct, the whistle call, like the wolves' howl, having more than one function.

In a letter, naturalist-conservationist Captain S. R. Banjeree from Shillong concludes from observations of dholes in the Garo Hills of Meghalaya, "when prey is sighted, the dog spotting the prey emits a whistling sound. At this, all members (of the hunting pack) carry out the attack simultaneously from all directions."†

Table 6.2 lists some of the sounds so far identified, and the various contexts in which they have been heard. The same sound may occur in different contexts, the sounds being indicators of the emotional state and intentions of the animal, rather than carrying specific information like human speech. The emotional "coloring"—the inflections and emphasis—collectively termed paralanguage that we add to the words we speak, and our nonverbal utterances, groans, screams, sighs, and so on, are analogous to these vocal signals of the whistling dog. A growl, for example, has

---

*For this reason they are perhaps best called whistling dogs rather than wild dogs (since many domestic dogs are feral or "wild dogs"). The Hindi and Tamil names of dhole or son kutta and chennai add little except local flavor, since translated they simply mean wild dog and red dog, respectively.

†Letter to A. J. T. Johnsingh, Dec. 14th 1978.

Table 6.2 *Vocal Repertoire of Wild Dogs*

*Sound Type**	*Context*	*Context (in Johnsingh's study)*	*Function*
Whine, Whimper	Friendly approach, greeting, food soliciting.	Food soliciting, when scented by presence but when failed to see me.	Reduce social distance and maintain proximity.
Growl	Aggressive	As soon as I was scented and seen, before chasing village dogs, during interactions between dholes at kills.	Increase social distance threat, warning/alarm.
Growl-bark	Aggressive	When I intruded suddenly and when pups saw tourists on an elephant 100 m from their den.	Increase social distance, threat, warning/alarm.
Scream	Human intrusion/fear.	When tiger or leopard was met (indirect evidence)	Warning alarm, distraction.
Repeated yak-yak (chatter)	Human intrusion/alarm	Unexpected human intrusion.	Warning alarm, threat /challenge.
Repeated whistle	In dense cover, traveling or coming together.	After an unsuccessful hunt or after the pack was dispersed by the arrival of van.	Assembly or contact seeking: contact maintaining.
Mixed yip-yack cackle.	Closing in on prey.	When human intervention surprised them while closing in on prey.	Socially facilitative reciprocal arousal?
Yapping-howls	Rare, at dusk.	Were not heard.	Communal activity—group uniting or territorial function?

*Single sounds may be mixed in successive combinations, e.g. growl-scream and yak-yak-scream, or in simultaneous combination (superimposition), e.g. growl and scream combined. Johnsingh also reported squeaks or whines from pups during play.
Johnsingh (1982) did not record the hysterical yapping-howl in concert at dawn or the wailing contact call at the end of the chase reported by other naturalists.

different meanings in different contexts. In one context, the dog may be giving a threat signal to a rival and, in another, the sound may serve as a warning to cubs that danger is near. Clearly, the animals must be aware of the different contexts within which they are communicating with a limited repertoire of signals. Such awareness of context, termed "metacommunication," lifts the level of communication above that of purely instinctual acts. This awareness is best exemplified by play, where one animal may rush at another in a full-blown attack and yet administer an inhibited bite. Its companion, on seeing the attack, rolls over ("attack me") or runs off ("chase me") and shows no defensive aggression because the "mood" has already established that "this is play." A prior signal, in the wild dogs a "play bow" and a "play face" expression, sets the mood and determines the context of the interaction.

Some sounds can occur at the same time, like a mixed growl and a scream together may signal fear (or warning) and the growl an intention to attack if you come any closer. Rapid sequential shifts from one sound to another (i.e. unmixed, growl then scream) can have the same effect. These possible variations, with changes in sound intensity and in frequency of repetition, form the basis of a rich "emotional language."

The sounds' effectiveness is enhanced not only by metacommunicative awareness but also by their being linked with a number of facial expressions and body and tail postures (see Table 6.3). These visual signs too can vary in intensity, and some signals can be superimposed or simultaneously combined just like certain sounds. For example, an open mouth (Figure 6.10) combines the submissive grin with a defensive tooth display, either of which can also occur separately in other contexts. This expression conveys two different emotions at the same time and its effectiveness is enhanced further by changes in the positions of the tail and ears and the angle of lean of the body (or distribution of weight on fore and hind feet).

Johnsingh includes the following additional observations on dhole communication in his thesis (1979), noting that:

> a) Dholes may wag tails like domestic dogs, while pleasing others. Pups, nursing bitch and adults showing active submission wagged tails. Bent Ear while resting on the ground near the den wagged tail when another dhole approached. From the hide the body posture of the other dhole could not be seen. b) Submissive rolling was seen two times, once while playing and the other time while eating, and when the opponent growled keeping the ears flat. c) Dhole while attacking a prey and when chasing crows keep their tails arched. d) A subadult in Sigur stalked me once (keeping the limbs stiff and extending its muzzle) in the manner of African wild dogs.

Communication in the wild dog is clearly based in the present. It is a context-bound, emotional language which expresses what the animal feels as well as its intentions (to attack, play, etc.). A third channel of communica-

Table 6.3 *Body Language Repertoire of the Wild Dog*

Facial Expression	*Other Signs*	*Context*
Horizontal lip retraction, (submissive grin)	Tail wagged in low position or tucked.* Licking with tongue.	Friendly greeting, submission.
Mouth open with lips back (play face).	Tail may be vertical; play bow. Exaggerated approach withdrawal.	Playful intentions.
Lips forward in pucker or snarl.	Erection of hair on back, tail horizontal or vertical.	Threat, offensive aggression.
Lips back horizontally.	Tail tucked, ears flattened.	Fear, defensive approach.

*May urinate submissively, or with excitement.

tion is smell. We have no evidence of wild dogs using urine to mark their routes or territories, as is typical for many other canids. They do deposit their feces in very obvious places, such as at the intersection of two trails or at the point where a trail leads to a clearing in the scrub. Dozens of feces of different "ages" (as judged by different degrees of weathering and breaking up from the feet of other passing animals) may be found at a dunging site. Repeated visits to such sites and the discovery of new sites on various routes of travel revealed that the dogs clearly prefer to defecate in unison and have a predilection for places they have used before.

No consistent pattern of marking in relation to place or time was noted, however. A territorial function is unlikely because several heavily marked areas were well within the dogs' hunting ranges rather than toward the periphery. Since no adult feces were ever found near dens, and dunging sites were often marked several hundred yards from the dens by the hunting parties moving out on the morning hunt, these dunging sites are probably no more than communal latrines. Further studies are needed to evaluate this deduction, since the impressive latrines could serve some communicatory or social function. Johnsingh observed six to nine dogs in a pack of fifteen defecating together at communal scat sites and has proposed that this group behavior may help the pack determine how recently the area has been hunted. At least the latrines do attest to the dogs' den hygiene and to their group-coordinated tendency of together doing the same thing at the same time and place.

No signs of scraping with the feet after urinating or defecating were

Figure 6.10 Features of adult male whistling dog; submissive-defensive gape (in lower photo) toward author (captive animal, Paris Zoo).

found.* Such scraping in domestic dogs, wolves, and coyotes may have some territorial marking function and may occur in the vicinity of a rival pack or individual animals. As in other canids, the wild dog has a well-developed scent gland one-third down the upper surface of its tail. The sweet, musky odor of this gland is faint, like that of the wolf. Its purpose remains to be understood. Like other canids, wild dogs will roll in odoriferous materials, one observer reporting he saw one pack member rolling in the freshly voided feces of a companion.

*Keller (1973) reported seeing wild dogs marking each other with their own urine, some actually performing a "hand stand" on their forefeet, with the hind feet off the ground. Such social marking is not uncommon among mammals and probably serves to maintain a bond of familiarity or "olfactory kinship."

Figure 6.11 Vertical tail posture may indicate general arousal/excitation as well as being a social signal between dogs. (Photo courtesy of A. J. T. Johnsingh)

In addition to knowing the above "language" of visual and auditory signals, the wild dog must also be able to recognize the many sounds and fleeting visual cues of its prey and of the other inhabitants of the jungle (described in chapter 1) as well as a vast number of different odors. An animal born and raised in such an environment must be radically different in many ways from one raised, even with its own species entirely, in captivity. The breadth of experience necessary to survive and the amount of information a maturing wild dog must acquire early in life must be enormous. The question of protocultural traditions—of older dogs teaching younger generations where, when, and what to hunt, what to avoid and so on—warrants further study. Young wild dogs are extremely curious and, as a consequence, often foolhardy, being easy targets for hunters. This intense inquisitiveness reflects their readiness to explore their environment and to experience much during the formative months: exploratory behavior and intelligence are closely correlated.

A detailed study of the behavioral development of a litter of whistling dog cubs, even under conditions of captivity, would be extremely valuable. Burton (1941) noted that hand-reared cubs were quite aggressive, fighting frequently with each other until seven to eight months of age. After this age, there were no more fights and one dog was the recognized leader. Presumably, a dominance hierarchy had formed, but in the absence of adults the normalcy of this situation is questionable. Under normal conditions, the cubs would be subordinate to adults and the presence of the latter would undoubtedly influence the frequency of fighting and social relationships within the litter. Confinement in a small space alone could increase the probability of aggressive interactions.

Early one morning we saw three adult dogs play-fighting in the play area adjacent to the dens. The game lasted for twenty-five minutes and, as usual in canid play, consisted of two phases. The first phase involved chasing and ambush, each dog taking turns stalking and ambushing another who was the "prey." A small maze of tunnels in the long grass was used as hides, from which the "hunter" would leap at its prey (see Figure 4.4, chapter 4). The message that "this is still play" was frequently signaled with a play bow, high vertical leaps, exaggerated rushing forwards ("I'm going to get you"), or running away ("catch me if you can"). The second phase involved physical contact, the actions in this sequence being derived not from hunting but from fighting with other dogs. The dogs would approach with bushy tails stuck out horizontally, slightly arched, or even straight up, and would rise up on their hind legs facing each other. Face-to-face, they sparred, wrestling and pushing each other with extended forelegs in a fox-like manner. Then one would back down and run off only to be chased and hit side-on with a body slam from its companion's hip or shoulder. (This action may also be used in hunting.) Instead of running off, one dog might roll over and extend its legs to fend off its partner while the latter tried to bite its cheeks. Another action was to turn away rapidly as the partner tried to bite, using the hip or shoulder to block the attack and possibly knock down the playful adversary.

During his field study, Johnsingh (1979) also observed some play behavior and noted that:

> Play in pups involved jaw wrestling, ambushing, chasing, submissive rolling and face to face sparring while standing on hindlegs. Adult play behaviour included play soliciting, catch-me-if-you-can, homo or heterosexual mounting, chase and ambush. Twice adults climbed fallen trees. The tail was often kept arched. Adult play with the pups was mostly chase and ambush play over and in between rocks."

Another playful interaction was seen involving an entire hunting pack early one morning before making a kill. We watched four dogs chasing each other round and round a bush on the river bank; again, the actions of chasing and ambush, derived from hunting behavior, predominated. Wild dogs may engage in what Johnsingh (1979) interprets as a prehunt social ritual or mood synchronizing activity. He described this ritual as follows:

> Dholes were observed to perform this rite on only four occasions. In all four instances they rested and played on roads and on three occasions near

> their dropping sites (hunt 8 and 21). While resting they nuzzled each other, lay close to one another and rubbed their bodies against others. Active submission and tail wagging were common. Thrice they were seen walking on branches of fallen trees. Homo or heterosexual mounting was noted once. Chase and ambush were common. On one occasion this peaceful get together suddenly gave way to vigorous play which abruptly subsided and the pack started off to hunt. Vocalizations, at the climax of this behavior reported in African wild dogs and wolf were not heard.

What is the function of play? These dogs were relaxed and obviously enjoying themsleves. Reportedly, younger animals play to "practice" certain actions such as prey-chasing and fighting. There is little evidence to support this theory in canids (Fox 1971). Play is a way of coming together and staying together for a time, a means of reaffirming friendships and consolidating social bonds. It is stimulating too, not only physically but in the countless variations and repetitions of sequences that can be created. It is like dancing, where two or more animals sharing the same mood take up the same cadence for a while until one sets up a new rhythm, a different pace, a new variation to the game. Animals that play together stay together; it perhaps best exemplifies the unity of a family or clan and is one of the most beautiful things to observe in the wild. But the whistling dogs may not have many years left to play; the thousands of years of evolution, during which their behavior, social organization, and hunting patterns have become an integral and harmoniously balanced part of the ecosystem, may be erased forever.

If this book has so far told less about wild dogs than it might—for reasons already described—then let its omissions provoke and stimulate more inquiry into these remarkable and beautiful creatures. For me, these animals epitomize that elusive, indescribable quality of wildness, intelligent life, and relationship, embodied in the totality of the jungle wilderness. The whistling dog (as might a different creature for some other person) stimulated my scientific curiosity and drew me to India; it became my key to the jungle wilderness. Consequently, the ensuing chapters explore this kingdom, the nature of wildness, the relationships between man and nature, and also those aspects of human nature which affect man, animal, and wilderness alike.

Chapter 7

POLITICS, PEOPLE, AND THE JUNGLE

THE NATIVE PEOPLE

There are three tribes of people indigenous to the study area, and their diversity complements the variety of flora and fauna of the jungle. Closest to nature are the Kurumbas, who have three tribal divisions, the Jen, Mullu, and Vettu—forest gatherers, cultivators and bamboo cutters, and laborers and artisans, respectively. "Jen" means honey, and the Jen Kurumbas are an experienced jungle people. Their economy is based principally upon what they gather from the jungle, a tradition which has remained unchanged possibly since the origin of man as a hunter-gatherer. The Jen collect fruits and tubers for personal consumption as well as other forest produce, notably gallnut, soapnut, nutmeg, mosses and lichens, and discarded deer antlers, for cooperatives of the Forest Department. Skilled in jungle crafts, they are excellent trackers and mahouts, or elephant drivers. Interestingly, a mahout actually takes on the name of his elephant, a merging of identity between man and beast where the latter, used in forest logging work, may well outlive the former. Although regular employment is available to the Jen in the Forest Department, most of them prefer the freedom of the forest and seem incapable of changing their lifestyle for what others believe to be more progressive: their fulfillment is already complete in the material and spiritual benefits derived from the jungle. Religion is essentially vitalistic, reflected by an animistic belief in sacred spirit places, rocks, and trees. In fact, each tribal division of the Kurumbas has its own special deities through which the members communicate with spirits of their ancestors.

The second tribe is the Irulas *(irrula* meaning darkness). These people gather forest produce but, in addition, will cultivate the land. Irulas are also excellent trackers. Religious practices are comparably animistic and these people will often share in religious festivals with the Kurumbas.

The Paniyans comprise the third indigenous tribe, their name literally translating as "laborer." They are field workers and, until recently, were held in bondage by their landlords and transferred along with land like livestock. In the old days of tiger hunting, these people were respected for their courage in helping drive, net, and spear these dangerous but unfortunate creatures. The Paniyans are probably of an ancient aboriginal lineage, having negroid features—curly hair and flat, broad noses. They also worship animistic gods, but have less knowledge of the jungle and are probably even more limited by fear and superstition than the Irulas and Kurumbas.

A fourth tribal group, the Badagas, are not indigenous, being recently immigrated farmers and field workers. All these peoples are increasing in number, their land development programs increasing with them. These factors, together with continued destruction of jungle habitat, make a poor prognosis for the future of wildlife in the area. The closeness of all these tribal people with nature and their intimate knowledge of wild creatures is correlated with a high degree of animism and superstition which often confuse and confound their more applied knowledge of wild animals. Consequently, a rational attitude toward natural phenomena is often preempted by myth and superstition. For example, physical sickness will often be blamed on a spirit from some tree entering the body.

Yet, such illogical reasoning and the correlation of unrelated evidence while searching for cause and effect relationships is not exclusive to these tribal people. The attitudes of some of the more educated Indian and European hunters toward wild dogs and some of their interpretations in the field suggest a comparable deficit of insight and logic.

It is really incredible how an animal (or even a person), having gained a bad reputation, quickly becomes a scapegoat. Negative expectations surround it, as well as negative conclusions. Not only is the whistling dog blamed for the decline in chital numbers, but more recently it has been blamed for the decrease in large-antlered stags, since few impressive heads have been shot in recent years. In spite of finding no evidence in the field to support this notion, the regular hunter in this region does not consider an alternative hypothesis and sees this theory as a favorable reason to eradicate the wild dog completely. Seemingly, it is a general characteristic of the human mind to look for answers and jump to simple cause and effect correlations. How better the mind functions with a little introspection: if the hunter would look up the barrel of his gun instead of down it, he might see the situation differently.

Figure 7.1 The graziers and forest tribes of the Nilgiris are close to nature, as witnessed by these ancient and modern temples where the people revere the spirits of the jungle and of their own livestock. But their faith is not reflected in their actions, since their grazing practices, coupled with successive years of drought, are destroying the jungle.

HUNTING AND CONSERVATION

Following the practice of most hunting authorities, the Nilgiri Game Association, established at the turn of the century, places a minimum limit on the antler size of deer a hunter may shoot. A shikari may lose his license if an undersized stag is shot, and fawns and does may not be killed. This practice, motivated solely by the desire to bag a trophy animal and on protecting young male breeders not yet in prime, must be seriously questioned. When natural predators are reduced in numbers or are absent in a given area, the surplus young, aged, and sick members of a herd survive and add to the problem of nonproductive overgrazing. Imposing an antler size limit will not eliminate older stags whose antlers are "growing back." Clearly, more efficient game management should include not only annual population counts of indigenous game animals but also some provision for hunting fawns and does in certain years.

Nothing is known of the effects of removing prime stags from a herd. Their removal may not only be detrimental to the social organization of the herd, but may also negatively influence herd fertility. Older or inferior stags may breed less effectively in the prime males' absence. The visual displays of males of many species influence reproduction by effecting the endocrine systems of females (being courted) and other males (being competed against). The hunting and killing of prime stags might therefore have a slight effect on the entire herd which could be magnified significantly by the other detrimental factors in the habitat.

American Indians have told me of the gradual degeneration of white-tailed deer in some areas, not only in herd size but in physical stature and antler size of mature stags. This degeneration has taken place over the years and the Indians attribute it to trophy hunting. Rather than killing for food on a more random basis and taking the easiest, as is the natural pattern of the wild dog, wolf, and native hunter, the trophy hunter is highly selective and crops only one small segment of the population—the best males. Over several generations it is believed that this pattern of hunting affects the vigor and stature not only of the males but of the entire herd. Acting like a form of natural selection, selective culling of large males is bound to cause some change in the composition and physical characteristics of the herd. This result would seem to hold true in other parts of the world also, and is most likely why large chital stags are less numerous in the study area than a few years ago. Hunters naturally have a vested interest in conservation and

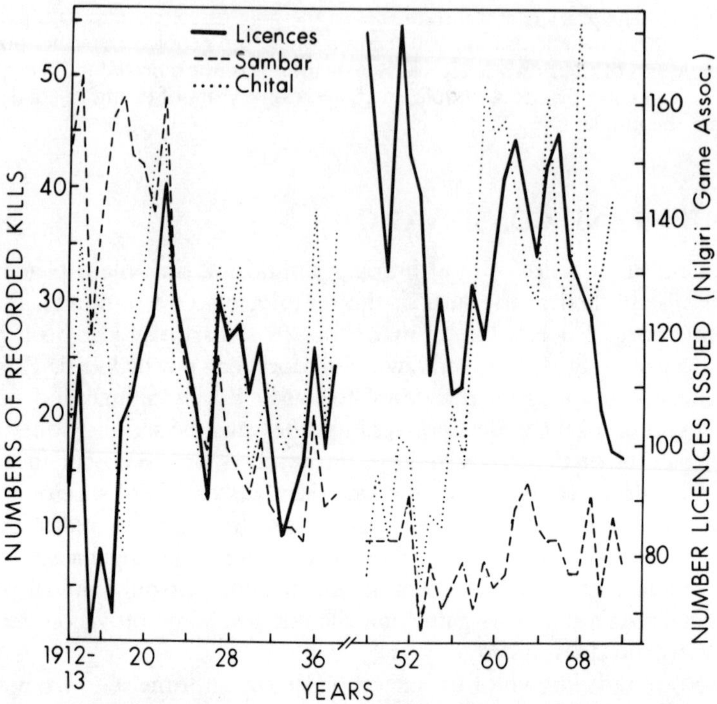

Figure 7.2 After the 1920s the sambar population shows a gradual decline (reflected in these data of recorded kills by Nilgiri Game Association). This decline may have been due to epidemic disease and relative overkill at this time, i.e. too many hunting licenses issued. An excessive number of issued licenses in the early 1950s caused further declines in both species of deer, the chital later recovering to where it was more numerous than sambar.

should become more acutely aware of the possible long-term consequences of their selective predatory patterns.

The Nilgiri Game Association's records of the numbers of animals killed each year provide an invaluable retrospective view of the changes in the ecosystem over the past fifty years (see Figures 7.2, 7.3, and 7.4). For all species designated as game, the following regulations evolved over the years: for all game (except sambar and pig) there is a closed season from July 1 to October 31. Sambar are shy, primarily nocturnal ungulates, and are difficult to bag. At one time they were abundant, judging from the number of recorded kills in the early 1900s. Their gradual decline over the years must indicate that the ecosystem is undergoing a steady change. A season closed to all hunting during their breeding period may help. After 1932, a lower limit on antler size was allowed (from 30 inches to 28 inches) to reduce the number of older stags whose racks were "growing back," and the bag per license was reduced from three to two. The constant killing of the best stags was thought to be associated with the smaller size of

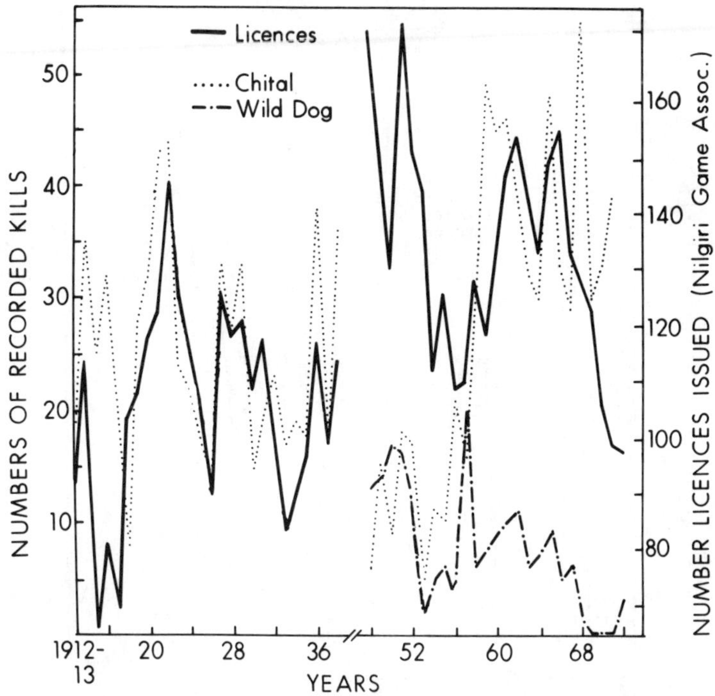

Figure 7.3 Number of hunting licenses issued and of chital killed show close correlation over the years, except during a possible overkill period in the early 1950s. The wild dog population shows a gradual decline between 1956 and 1968.

antlers. Fortunately there is now a closed season on the chital, or spotted deer, and a 20-inch minimum rack width determines which stags may be taken.

According to one knowledgeable former member of the Nilgiri Game Association, the incredible decline in the sambar population coincides with the construction of a large dam in the area. From the late 1930s to the late 1940s, these animals were shot by Canadian construction workers. One man alone amassed a collection of fifty-seven sambar stag heads or "racks." The impact of such slaughter on a massive scale resulted in a drastic change in the ungulate population, with chital becoming more numerous and the sambar never recovering its former status. Similarly, the decline of the leopard population during this time period, so apparent on the graph (see Figure 7.4), may be attributed to excessive and uncontrolled hunting by these foreign visitors who, unlike the local inhabitants, killed for sport rather than for food and were equipped with good hunting rifles, a commodity no village could afford.

This finding serves as an important reminder for conservation and wildlife management: construction crews working in wilderness areas could,

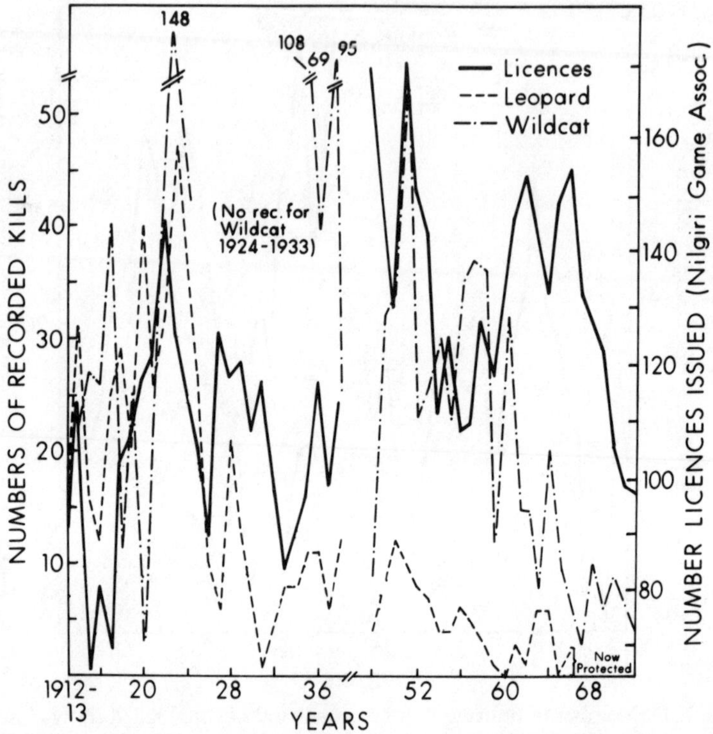

Figure 7.4 Great numbers of jungle wild cat were killed between 1912 and 1936, showing a marked decline in post–World War II years. The leopard, at a low level in the early 1950s, was drastically overkilled, but is fortunately now protected. Tiger, always rare in the Nilgiris, after the overkill of the early 1920s, is also protected.

if given guns, create havoc with the indigenous fauna, irrespective of the potentially deleterious consequences of the construction project itself. But to propose gun control laws even within the United States meets with strong opposition from powerful lobbies such as the National Rifle Association! There is a difference between a man's right to own a firearm and his right to kill any living thing with it. Unfortunately, the two are interdependent and are rarely separated by responsible thought and action.

The bison, or gaur, which graze at higher altitudes, were almost exterminated by an epidemic of rinderpest in 1909 and again in the late 1960s; no hunting of gaur was permitted until 1917. These animals are relatively rare in the Nilgiris, although they were once a favorite trophy animal. They are now regaining their numbers after more rigorous control and vaccination of domestic cattle.

A bounty is paid in India for the following species designated "vermin"—wild dog, wild cat, otter, red and gray mongoose, eagle owl and harriers, sparrow hawk and crow-pheasant. The bounty for a wild dog used to be twenty-five rupees (Rs), but since 1968 has been reduced by ten Rs. Forty years ago, twenty-five Rs was an attractive incentive making wild dog hunting a profitable activity. Pythian-Adams (1939) noted that the Game Association paid a higher bounty on the wild dog than in other areas, so there was a significant importation of skins by bounty claimants. The bounty was thus reduced from twenty-five Rs to twenty Rs in 1926. Many people gained a livelihood this way, some ninety-three bounties being paid out in 1922–1923 in contrast to an average of twenty-eight per year from 1933–1938 after the bounty was reduced.

Seemingly, the bounty management policy is aimed at insuring the availability of more small game for the hunter, such as jungle cock and hen, partridge, spurfowl, snipe, quail, woodcock, wood pigeon, hare, mouse-deer, and muntjac (jungle barking deer). This policy should be seriously questioned, and animals designated as vermin should not be shot indiscriminately. Rather, killing should only be done to regulate their numbers after a rigorous evaluation of their density and impact on the ecosystem. The need to protect these "nongame" species is still not fully appreciated even by fish and game authorities in many states in the United States.

Over the last twenty years there has been a rapid depletion of large cats in Southeast Asia; as a consequence of hunting, poaching and poisoning, the cheetah, lion, tiger, and leopard are either extinct or extremely rare in habitats where they were once abundant. Efforts to conserve the few remaining lions and tigers in India are offset by local politics, by primitive, inefficient, and shortsighted traditional agricultural practices, and by continued poaching and poisoning.

For these large predators, a bag limit of one tiger and leopard per license was made in 1960–1961, but this regulation was too late and too limited to save the big cats. It was a shortsighted compromise and by 1965 both tiger and panther (or leopard) were protected. In our study area of over 25 square miles we found no signs of tiger (although tiger was reported in the vicinity) and only tracks of two leopards. As far back as the 1930s, the increase in the wild pig population was recognized as being due to a decline in the panther population normally regulating the pigs' numbers, but not until 1965 was any action taken to save this carnivore. In 1960–1961 a bag limit of one was also set for the sloth bear, now less common in the area.

Davidar's report (1966) for the Nilgiris Game Association (personal communication) emphasizes that the elephant population is markedly increasing and that an average of six people are now killed annually by these animals. Elephants are now forced to raid plantations and villagers' fields because they are too numerous relative to their own shrinking and overexploited habitat. In the Mudumalai region of the Nilgiris, and undoubtedly in other areas, the elephant plays an intriguing and important role. The fear people feel toward elephants is perhaps a major deterrent to poaching in the jungle; the elephant is undoubtedly a more effective "game warden" than the forest officers who rarely venture from their vehicles off the jungle roads, no doubt for the same reasons as most poachers.

Like the Indian lion and tiger, the elephant is protected, and penalties for ivory poaching are severe. The elephant is regarded as a "sacred cow" by many and this attitude, coupled with its protection from poaching, is a major obstacle in its management. As in some parts of Africa, an overabundance of Indian elephants can rapidly destroy the habitat, especially in the dry season or in prolonged periods of drought. Trees are stripped and often killed and the vegetation may take years to recover. Lone tuskers and herds alike will keep on the move, going from one browsing area to another or working among the bamboo stands along rivers. With the expansion of plantations, such as wattle and eucalyptus, and of farm crops, there is a corresponding decrease in available habitat for the elephant (and for other herbivores, including domestic cattle) so that what land is available may soon be overgrazed and ultimately destroyed beyond recovery. Replanting indigenous trees and shrubs, regular census taking of the elephant population, and rigorous control of their numbers is urgently needed to preserve and restore what natural habitat is left.

Hungry elephants frequently break into plantations and crop enclosures and are driven off with firecrackers and anything that will make a loud noise; many show scars on their bodies from shotgun wounds. As a consequence of this kind of interaction with man, the Indian elephant in rural areas is unpredictable and often extremely dangerous. In the Nilgiris, at least four people and often as many as twenty are killed annually by elephants.

The following essay topic, prepared by the Nilgiri Wildlife Association for college students to compete for a prize, clearly reveals the change in thought and future action taking place today: "Wildlife is no mere heritage to be put on a pedestal or made a fetish of. Even the rarest animal is a renewable resource to be fostered, husbanded and harvested in due time." The general consensus is very much in favor of this view. But the earlier religious and philosophical attitude toward nature could make a better foundation for conservation efforts than this western anthropocentric attitude of management and exploitation.

While consistent study of the whistling dogs provided few sightings, a chance encounter by a person not studying them daily could lead to the general but erroneous impression they are abundant. One person seeing an assembled clan of twenty to forty dogs would immediately conclude the "packs" must be very large (while in fact, these are not hunting "packs" *per se* but a social aggregation of hunting units). Forest officials draw illogical conclusions from such reported sightings by visitors (and also from their own chance observations). From such impressions the wild dog population is estimated. The estimate is immediately magnified (illogically) if the deer population (calculated with some accuracy from year to year) is on the decline. The wild dogs are "blamed," the estimated population automatically increases, and there is talk of reducing their numbers. Hunters in shooting areas adjacent to sanctuaries are encouraged to shoot wild dogs on sight.

Even the District Forest Officer (in charge of several sanctuaries in the area) was less concerned about the impact of domestic dogs on wildlife than the impact of the whistling dogs. He told me he hoped that, within the sanctuaries, rabies or distemper might wipe out the chennai for a while and so help the deer recover their numbers.

Similarly in the United States, less educated people blame a declining deer or moose population on wolves, just as the whistling dog is blamed when Indian deer herds seem smaller. Inadequate food (as in a mature forest with no underbrush) or poor regulation of the number of hunting licenses issued, combined with a couple of severe winters (or dry season in India), are usually the causes. Diseases from domestic livestock may also be implicated. But what if none of these factors caused the population decline—can we then rightly blame the wolf or the whistling dog and cull the predators so the deer will increase? If the predator population is high, it should be left to regulate itself naturally—as it has done for hundreds of thousands of years. We do not have sufficient knowledge yet to intervene when such population declines occur; they may be part of much longer, interrelated cycles lasting many years, essential patterns maintaining the balance and diversity of a given ecosystem. Hunting is the one overriding reason man so often intervenes (a minority interest at that) but this is no valid reason at all, either ethically or ecologically.

DOMESTIC LIVESTOCK AND WILDLIFE CONSERVATION

Forest officials are obliged by local politics to allow cattle to graze in the sanctuary and are informed they must turn a blind eye to illegal hunting by higher-ranking government officials and local, influential people. Trees may be felled to provide food for hungry cattle, especially in drought-stricken areas (which includes most of India), and graziers will burn vast areas of land to improve the grass for their cattle. Such burning may selectively kill many plants important to the ecosystem and vital for the health of nongrazing herbivores such as browsing deer (chital and sambar). Some graziers are mobile, shifting their pattis several miles when their cattle have grazed out one area of the jungle. Tradition rules that male calves may not be killed but "left to nature." These animals are of no use because cattle are kept only for their milk. Old nonproducing milk cows are similarly never destroyed. So, a great portion of a herd may be nonproductive but continue to graze—a ludicrous economic situation adding further to habitat destruction.

Figure 7.5 Temporary quarters may be set up by graziers (upper figure), while more permanent pattis are the rule. Cattle and calves are corralled between 5:00 P.M. and 8:30–9:00 A.M., thus fortuitously avoiding the early-morning and late-evening hunting periods of wild dogs in the hot, dry season.

Paradoxically, such "unwanted" cattle might be left out at night and not corralled with the herd in the hopes that a tiger or leopard might kill them. If the nonproductive cattle survive, they join and graze with the herd the next day. And yet, if a predator killed a milk cow, poison bait would be put out for the killer. Clearly, predators are supposed to discriminate between productive and nonproductive stock. The easy availability of the latter must have given more than one cat a taste for cattle: indeed, the predators were encouraged to take such cattle.

The contrast between the areas where domestic cattle grazed and similar inaccessible land within the adjacent Mudumali sanctuary where cattle grazing was restricted was strikingly apparent after the rains. In spite of the rains, areas grazed by cattle had a mown, parklike appearance. The impact of such overgrazing is, therefore, apparent even in the rainy season, emphasizing how destructive and self-limiting this free-range grazing of scrub cattle is. In the dry season, with even less forage available, a herd of several hundred cattle will only yield a few liters of milk.

The only positive step in wildlife protection so far has been the vaccination of all domestic cattle for rinderpest and hoof and mouth disease. Local politics essentially prevents the Forest Department from banning all cattle grazing in wildlife sanctuary areas. Until such a ban can be effectively accomplished, wildlife sanctuaries in India will not be optimal habitats for indigenous wild species. At least the disease control program for domestic cattle has helped restore the gaur or Indian bison, decimated a few years ago by a rinderpest epidemic. Similarly, more rigorous control of indiscriminate poisoning (especially with the pesticide Follidol) of the carcasses of lost livestock has helped restore the panther, tiger, and hyena populations in some areas.

A first step in improving the situation in relation to destruction of natural habitat by grazing cattle would be to reduce the number of cattle and improve their quality by developing more productive hybrid varieties. Scrub (Brahmini) cattle, crossbred with Channel Isle, holstein, or Brown Swiss cattle, would combine good heat resistance with high productivity (yielding twelve to thirty liters of milk per day compared to 4 or less liters by native cattle). A buffalo requires more fodder to produce the same quantity of milk as a Brahmini cow, and the former is of questionable utility except for work.

With fewer, higher-yield cattle the problem of overgrazing would be reduced. Overgrazing could be reduced further by keeping the milk cows in feed-lot corrals. Cattle would then have to expend less energy roaming in search of suitable food. Farmers should be encouraged to grow some crops for cattle food supplements and make more hay and silage for the dry seasons. The expense of providing water for semiconfined animals could be subsidized by savings in forest resources and also by profits from more intensive dairy practices.

The National Dairy Research Institute of India has been promoting these ideas in spite of local tradition and opposition. The Institute has demonstrated that a one-acre plot can give a farmer an adequate monthly income with three high-yield hybrid cows and crops grown in a recommended rotation throughout the year. One cause of resistance to the adoption of this project has been the humpless bullocks the crossbreeding produces. Brahmini bullocks have humps, which allegedly give them strength and help hold the yoke for draft work. The usefulness of hybrid bullocks as draft animals remains to be demonstrated to village skeptics.

A comparative analysis of draft animals and a tractor, comparing the cost of fodder and gasoline and the range of farming operations achievable by each, places the latter superior in many respects. Even the initial price of the tractor can be balanced against the cost of raising a calf to working maturity. Man-hours are also saved by using a tractor, which could be purchased and shared by a group of cooperative farmers and supported by a low-interest loan from the government. Draft animals will continue to play an important role for other purposes, however.

Draft cattle and large herds of low-producing cows are slow to become a thing of the past, although the problems are well recognized. This situation must be rectified as soon as possible not only to support an ever-growing populace, but also to reduce the devastation of forest and jungle.

In summary, Indian cattle graziers' livelihoods are at a subsistence level most of the year (except after the rains) and are one of the most destructive influences on the ecosystem. In addition, the cattle have frequently infected wild ungulates with brucellosis, rinderpest, and hoof and mouth disease—epidemics to which the wild ungulates are extremely susceptible. It takes many years for the natural population to reach normal numbers and ecoequilibrium.

Hungry villagers will poach whatever they can, encourage their pi-dogs to hunt (see chapter 9), and cut trees for firewood and fodder for their stock. Rural communities, following inefficient agricultural practices and faced with repeated crop failures and livestock mortalities from drought and disease, must contend with an ever-increasing population and threat of famine. How can natural ecosystems be preserved and sanctuaries maintained under such conditions, where dire human need supercedes all animal rights?

FIREWOOD AND COW DUNG

The increasing destruction of natural habitat and depletion of resources by rural communities is of critical concern for countries such as India. Ecodestructive influences spurred by an ever-growing population may very soon destroy the environment to the point of no return. The deleterious effects of forest clearing for agriculture and overgrazing by domestic livestock are further intensified by removal of much of the cow dung (from the overnight corrals), taken away and sold as fuel. This vital component of the ecological cycle is being burned in the cooking fires of the villages and small towns of India. Another resource—firewood—is also stripped from the jungle because of kerosene's high cost.*

An ever-expanding zone of habitat destruction typifies the village environs. With such deforestation, a secondary effect—land erosion—is inevitable. Such erosion in the Nilgiris is evident but, as yet, is sufficiently limited so that remedial and control measures would be effective if instigated soon. The planting of rapidly growing trees, such as eucalyptus and adaptable varieties of pine, to provide a communal source of fuel would be advantageous. This step would merely be a stopgap, however, if measures to ensure population stability and more efficient and ecologically sound farming practices are not effected. The overwhelming need for reforestation and soil conservation in developing nations is often overlooked when the people's only source of fuel is wood from the forests and dung from their cattle.

---

*For an excellent review, see E. P. Eckholm, *Losing Ground* (New York: Norton, 1976).

Chapter 8

THE PLACE AND VALUE OF THE WHISTLING DOG

In the absence of larger predators, the importance of the whistling dog in maintaining chital herd quality cannot be overstated. An ecosystem in which there are no natural predators is analogous to a rose that is never pruned: uncontrolled, prey-species will rapidly grow in a self-destroying profusion of smaller and weaker generations until all that is left is overgrazed scrubland with neither flower nor flesh. The wild dog is an extremely adaptable predator since its numbers can be regulated naturally in relation to prey availability much more rapidly than the large cats' population. Whereas a given habitat could at best support only a few large solitary hunters such as the tiger and lion, the wild dogs are more flexible in that they can take a wider variety of prey as food, and can hunt singly or in large packs to secure much larger prey. The wild dogs can also range much farther for food than can the big cats. With communal care of young and with more than one female producing young in a pack (when game is plentiful), the dogs can quickly restore their numbers after a population decline, whereas the tiger and leopard must raise their litters without any support from conspecifics. In a relatively stable ecosystem, the big cats are well adapted, but once instability sets in, their survival will be threatened. The more versatile whistling dog, able to adjust rapidly to the many disruptive factors described in the previous chapters, are still able to survive while the big cats are now facing extinction.

With the absence or extreme rarity of tiger and panther in Indian ecosystems, the role of the wild dog in maintaining the balance of nature is now of critical importance. If the dogs are exterminated, the quality of game will suffer, since human hunting in selecting "trophy" stags is, as we have discussed, of little value in terms of herd management. If accurate head counts of game animals cannot be made annually, then we cannot hope to "manage" the wildlife in the absence of a natural predator. The wild dog must therefore be protected, because in its absence the ecosystem will certainly collapse. Any ecosystem without an apex predator at the top of the food chain is in disequilibrium and will ultimately disintegrate. With highly efficient management, man might be able to stave off this ultimate catastrophe. In the absence of adequate knowledge and practice, however, he had better leave this role to the natural predators who have been maintaining equilibrium for hundreds of thousands of years.

Many blame the whistling dog for the periodic declines in chital and sambar, but it is more likely that diseases from and competition with domestic livestock are to blame. The wild dog population itself fluctuates from year to year: in a year abundant with game more cubs will survive and, in the following year, their predation will be much heavier. If this second year is a bad year for the deer, because of disease or lack of food-resources, then the heavier predation by wild dogs will reduce their numbers considerably. The year after, however, mortalities in the wild dog population would be high because of starvation and, ultimately, the prey-predator population will reach a temporary equilibrium again. All natural systems oscillate, and the fluctuations in deer and wild dog populations often disturb the ill-informed and shortsighted hunter and wildlife manager. The same holds true in the United States and Canada where sporadic wolf-extermination programs are instigated to "protect" deer, moose, and caribou, ostensibly from the wolves, but in reality for the hunters. Such are the politics of extinction.

This natural prey/predator cycle can be seriously altered by domestic livestock. Wild dogs prefer to kill chital and sambar, but in years when natural prey is scarce, they will turn to domestic stock. With such an alternative food source the wild dog population may not be regulated naturally, thus posing a serious management problem. This problem exemplifies the critical interface between man and nature: either the wild dogs and the whole ecosystem must be destroyed or drastically modified by systematic shooting and poisoning, or the cattle must go. Neither alternative is desirable, but, unfortunately, the former alternative may soon be inadvertently adopted.

Some patti owners accept predation by wild dogs with equanimity, knowing that after a year or two of considerable loss, few stock, if any, will be killed in other years. The wild dogs will not move out of the area; they will return to killing their favored natural prey whose numbers have increased once more. Perhaps this attitude is the best way for man to live with

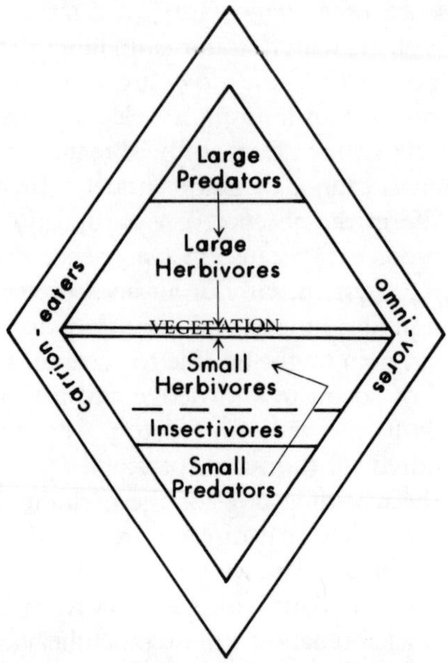

Figure 8.1 Schema of food-chain interrelationships of different species in a complex ecosystem like the jungle. Specific carrion eaters include vulture, hyena, jackal, and crow; omnivores—wild pig, porcupine, and sloth bear; small predators—fox, mongoose, jungle cat, and jackal; insectivores—fox and mongoose; small herbivores—muntjac, hare, and macaque monkey; large herbivores—elephant, chital, and sambar; large predators—leopard and whistling dog.

nature—to not always expect a one hundred percent return profit from livestock each year (see Table 8.1).

Other patti owners will poison a calf or cow killed by wild dogs with the hope of destroying the predators, but the whistling dogs rarely, if ever, return to a kill. The poison, instead, kills the carrion eaters, notably the jackal, crow, vulture, and village pi-dogs. The striped hyena is less common in the Nilgiris than in the past, probably both because of this custom and as a consequence of direct competition for carrion with semiferal pi-dogs, more numerous than indigenous jackals. These dogs, who have occasionally been seen hunting with whistling dogs when the wild dog population is low, undoubtedly contribute significantly to the disruption of the natural ecosystem. They not only compete with natural carrion eaters but also carry mange and distemper, the latter disease a suspected cause of high mortalities in the wild dog population in some years.

Hopefully, if more sanctuaries are set aside and existing ones enlarged and managed more rigorously in regard to poaching, poisoning, and grazing domestic cattle, a few of the remaining big cat populations may be saved and the endangered wild dog packs protected. In some places, these predators might also be effectively reintroduced.

Table 8.1

Patti	*Number of Cattle*	*Annual Average Loss*	*Recent Kills* (Dec.-Feb)*	*Worst Months*
Odaikarai	300	10	0	May
Thanatty	200	12	0	May October
Chembanatham	600	40	4	December, January, April
Chokanalli	250	5	2	November, December
Arakadavu	700	50	2	November, December, May
Mylar	200	25	2	December
Malaipuram	400	60	4	November, December
Total	2650	202	14	

*Kills reported just prior to study period in 1974.

Extermination of the whistling dog will only ensure that the ecosystem will degenerate even more rapidly than it has to date. When this animal can be appreciated for what it is, and its role in nature understood and valued, the ecosystem will have a better chance of being conserved. The same is true for any wild animal, since each has a place in the scheme of things and must be appreciated and valued in its own right, independent of any human value. Until recently, the wolf in North America and hyena and wild dog *(Lycaon pictus)* in Africa were much maligned by hunters, wildlife managers, and conservationists alike. The predator competes with the human hunter: both are after the same prey. In addition, the predator's often slow and bloody method of killing its prey, and even eating it before it is dead, adds little to its favor in the eyes of the general public.

Krishnan (1972) notes that wild dogs rarely take domestic cattle and have never been known to attack man. He feels that the anthropocentric antagonism toward the animals originates because they are seen as rivals by the hunter, disturbing "man's game." They are invariably regarded as vicious killers of deer which are so defenseless and esthetically pleasing. More than one naturalist has attempted to rescue a fawn from a pack of wild dogs, even to the point of shooting at the dogs.

Phythian-Adams, if alive today, might be more enlightened and would surely retract the following statement from one of his papers (1949): "The epithet 'a perfect swine' may with every justification be applied to the wild dog whose nature and habits may be summed up in a single word—bloody." He urged that the dogs be recognized as jungle pests and destroyers of game and that they should be eliminated, concluding with the statement: "Except for his handsome appearance, the wild dog has not a single redeeming feature, and no effort, fair or foul, should be spared to destroy these pests of the jungle" (1949). Even more recently, E. P. Gee (1964), a noted Indian naturalist, wrote that the wild dogs should be kept to a minimum, especially if deer are becoming scarce. Yet the wild dogs have hunted deer for thousands of years and if the prey are becoming scarce, there must be some other factor effecting the deer directly or causing an abnormal increase in wild dog population. The ethics and ecological soundness of wildlife "management" interventions, such as killing a supposed excess of wild dogs (or wolves, as has occurred in Alaska), are still open to serious doubt. The value of such practices, other than to reduce the competition between man and predator for the same game "resource," is questionable.

Gee concludes that wild dogs "do keep deer on the move and so favorite grazing areas do not become over-grazed and therefore impoverished. Otherwise there is very little to be said in favour of these animals." There is no evidence whatsoever to support this statement. Like many naturalists, he draws illogical conclusions and inferences from correlating events not necessarily related. Deer will move in search of food and the wild dog may follow them—such is the observation. Gee's interpretation makes an illogical correlation between the movements of prey and predator, inferring that the former benefit by the latter moving them to fresh grazing areas.

It is regrettable that such influential and well-intended people are so ill informed and that they perpetrate such scientifically unsupported conclusions much to the detriment of nature. Often their intentions are quite the opposite. But their influence on public attitude is considerable, even to the point of affecting the perception and thinking of conservationists and government officers working directly in the field.

If, however, human rights and needs continue to press unabated—population growth continues and inefficient agricultural practices remain unchanged—there can be little hope for the wildlife of India and other countries that still have relatively pristine wilderness areas. Against the urgent needs of the people, wild animals have neither rights nor spokesmen, and, because animals cannot vote, no politician in favor of conservation will gain election.

If some of the people's needs can be linked with the need to conserve nature, the politician/conservationist may gain some support in the near future. How can this link be accomplished without presenting the argument (which leads to even further destruction on species and habitat) that wilderness and jungle must be conserved and properly managed as a resource for optimal exploitation by man? The only solution, as many see it, is to totally remove villages from designated sanctuary areas and relocate them where there are adequate buffer zones between protected habitat and land designated for agricultural and commercial development.

Without an understanding of the natural laws of ecosystems and of the possible consequences of interference, man will never live in harmony with nature; the destruction of the earth will continue unabated and at an accelerating rate. What then? Will we then seek alternatives: alternative methods of agriculture, other sources of energy, other life styles, and more rigorous control of population growth? Without reverence for all life, based upon sound knowledge, rational action, and a value system that transcends the shortsighted goals of immediate gratification of human needs, there can be no future. It will be too late. Time is running out. There is time now to create a new value system where reverence for all life unites mankind in a common bond of understanding and commitment. But perhaps, by tomorrow, it will be too late.

Chapter 9

PARIAH DOGS OF THE CITY AND JUNGLE

In contrast to the free-roaming and feral dogs I have studied in the United States, Fox *et al.* (1975)* the abundance of dogs in the cities and villages of India was quite remarkable. At least one dog could be found along every few hundred yards of street, among the teeming people, the small, open stalls, the wandering cattle, and the scratching chickens. By all accounts, most dogs were homeless, and yet, a percentage of them were in reasonable condition. (By reasonable, I imply absence of extensive skin infections, emaciation, or injury to one or more limbs.) Some children showed me their pet dogs in Srinigar (Kashmir), but few were kept as pets. Children and adults alike avoided any physical contact with either pet or stray dogs. Some cities, like Bombay, and large towns, such as Srinigar, have instituted dog-catching and destruction programs, poison bait (strychnine) being used most often.

The displays of submission toward conspecifics were more typically coyotelike than doglike in these pariah dogs, i.e. more *exaggerated* than in Western dogs I have observed. A clear back arch (coyote trait) was combined with the typically doglike lowered head, flattened ears, and submissive grin (horizontal retraction of the lips) or defensive snarl (as distinct

*Portions of this chapter are from M. W. Fox, *The Dog: Its Behavior and Domestication* (New York: Garland Press, 1978).

from a wide-open mouth gape as in the coyote). The more outward set of the erect ears, and upward curling tail make the pariah dog quite distinct from Western mongrel dogs, although similar, dingolike canids are seen in Mexico and South America as well as throughout Southeast Asia. Short coats and erect ears are characteristic in the pariah dogs, the most frequent color being brownish red or brownish yellow. Body size averages approximately 25–35 pounds, males tending to be larger than females. Piebald (black/white or brown/white) shaggy and medium-long coats, and occasional flopped or pendulous ears are frequently seen—possibly a more recent influence of imported (European) dogs on the indigenous pariah types.

Although an occasional dog was seen chasing children, most were passively submissive or fearful and avoided approach. Although obviously hungry and quite capable of killing free-roaming chickens, the pariah, or pi-dogs, ignored or avoided them, probably through conditioning. A dog was once seen chasing a chicken and was immediately stoned and frightened away. An all too common sight was that of children and even adults throwing sticks or rocks at these dogs for no apparent reason other than sheer enjoyment. But at least the pi-dogs have one day a year with some grace. On the Festival of Khichmauas, people throughout India will feed all dogs with the hope that the famous stray dog (Khichmauas) will visit their homes and bring good luck. Unfortunately, even this festival reflects superstitious self-interest rather than human compassion for other creatures.

As I had anticipated from the study of feral and free-roaming dogs in St. Louis (Fox *et al.* 1975), dogs in large cities were seen singly or, rarely, in pairs, because the limited abundance and wide distribution of food dictates a solitary, scavenging mode of existence. Thus, the environment profoundly influences social behavior of both man and dog. Feral dogs in the St. Louis ghetto were most active between 11:30 P.M. and 7:00 A.M. during the summer. It was thought that this activity pattern was related to active avoidance of human contact. Other free-roaming house dogs had a similar activity peak, plus two smaller peaks around 7:00–9:00 A.M. and 6:00–9:00 P.M. These smaller peak periods were associated with people letting dogs out in the morning, and with people and dogs socializing on the streets in the evening. Little activity was recorded between 11:00 A.M. and 5:00 P.M.: dogs and people alike avoided the heat of the day. The Indian pariah dogs had a pattern similar to these free-roaming house dogs, their activity patterns reflecting heat avoidance (most sleep from late morning until late afternoon) rather than avoidance of people *per se,* as was thought to be the case with the feral dogs of St. Louis. Thus, the Indian pariah dogs appear to be more human-oriented, and their activity pattern coincides with human activity on the streets, although they do obtain scraps of food during the day from storekeepers and passersby.

Smaller towns and villages tend to be more supportive of both home-

Figure 9.1 Indian pariah or pi-dogs, following a bitch in estrus (above), and a bitch with litter (below). Note emaciated condition.

less people and dogs. Beggars make their rounds and shop owners give them a few cents or food not sold at the end of the day. Butcher shops and even bakeries can be identified at a distance since usually one or more dogs are close by, waiting for any scraps. The store owners disclaim the dogs and say they are strays. Perhaps feeding them in this way make the dogs tolerant, patient, and relatively social, thus unlikely to attack anyone unless rabid.

These customs are reminiscent of Dr. Hans Kruuk's report (1972) of wild hyenas coming into Ethiopian villages to be fed at night. Local people felt this action would keep them on friendly terms with the hyena pack, which might otherwise attack and carry off livestock and children. Pariah dogs, however, are not deliberately fed for that purpose, but rather because they are hungry.

Time did not permit a detailed study of these feral dogs; questions as to their social relationships and territoriality (adhering to one particular street zone and keeping strange dogs away) remain to be answered. A territorial system may exist, since the same feral dogs were seen in the same locales over several days. Eight dogs were seen chasing off another dog in an open area just outside of one village.

No less than eight dogs—four females, two males, and two young male dogs—were seen resting together in the shade in the large town of Srinigar, Kashmir. The largest male and female were clearly the most dominant. When disturbed, this aggregation (rather than pack) split up into five and two, the dominant female disappearing at least temporarily. Local people said the animals were all strays and, considering their poor physical condition in contrast to the few pet dogs that were seen, this statement was probably true. This same group of dogs was seen together several days later in the same area, suggesting at least a temporary pack formation.

In St. Louis, such large aggregations of dogs were only seen when a female was in heat and a group of males trailed her (Fox *et al.* 1975). A mixed group (of different ages and sexes) with no female in heat attests to the social capacity of the pariah dog. More research is needed on these dogs' social relationships, although any such research will undoubtedly be frustrated by the high mortality rates of the subjects under study.

The pariah dog remains an integral part of urban and rural life in India. The animal serves no specific function for man other than as a recycling agent in the biodegradation of garbage, food wastes, human excrement, and other organic materials. Its epidemiological significance in terms of diseases transmissible to man and as a passage-agent in reducing the virility of organisms pathogenic to man remain to be evaluated.* Although they are

*An estimated 120,000 Indian people per year seek treatment for dogbite wounds, especially in more crowded urban areas. Human deaths from rabies are common and often reach reportedly epidemic proportions, indicating that the feral and free-roaming dog problem in India is serious and that the most stringent measures are needed to control it. Collecting and humanely destroying ownerless dogs would certainly be of benefit to both the human and animal population. Even house pets are unsafe because of the potential exposure to rabid dogs, and rabies vaccine for pets is not readily available in most areas. Some municipalities use strychnine in bait to kill pariah dogs while others use dilapidated or improperly operated electrocution devices. Hopefully, more humane methods of population control will soon be funded and put into action.

rarely if ever touched, their excrement and urine in the dry season become a potential health hazard in the dusty streets. But this material may, through passage, provide people indirectly with oral inoculation of innumerable weakened microorganisms at an early age.*

It would be more aesthetic to remove the homeless dogs from the streets of India, but without changing other sanitary and social habits that measure could be ecologically (and epidemiologically) disastrous. You can only pity the deprived animals and homeless street people and admire the will to survive.

Nothing is wasted in this urban ecosystem, in so many ways like the true jungle-survival of the fittest with food-chain (and socioeconomic) interdependence. But, unlike the jungle where life is no less abundant, urban life is too often supported marginally and suffering is prolonged. The more merciful (but not necessarily humane) early end to the sick and less vigorous maintains the quality of life under natural conditions. Even so, only the healthier pariah dogs survive to maturity and reproduce. In comparison to the wild dogs of the jungle, however, few pariah dogs have the same health and vitality. Where the environment is harsh, only the healthy survive; but, when environmental selection pressures are less rigorous because of protective buffering (by man), many that would normally die somehow continue to exist and, so, to suffer in varying degrees.

Humanitarian goodwill may often perpetuate such suffering. Perhaps the healthiest animal and human communities are those with neither medicine nor misguided humanitarian benevolence for the sick and underprivileged. But then, what happens to altruism and reverence for life?

To feed the pariah dogs is to ensure continued generations of marginal subsistence and further suffering; the same is true for mankind. This fact exemplifies the dilemma between moral values and ecologically sound ethical conduct. Reverence for life is tested critically under such circumstances. You cannot ignore nor neglect what lives and needs and loves, in spite of the consequences.

The altruism of a chunky male pariah dog meticulously grooming a sick and emaciated bitch (who was not in heat) remains with me. He even reached out and snapped at the flies that buzzed around her face, and when thrown a few scraps of bread by a street vendor, he sat back and allowed her to devour every crumb.

Some fifty years ago, Mahatma Gandhi made this poignant statement which, like much of what he said and did, is highly relevant today:

> . . . it is an insult to a starving dog to throw a crumb at him [Presumably because its life of suffering will only be prolonged]. Roving dogs do not indicate the civilization or compassion of the society; they betray on the contrary the ignorance and lethargy of its members.

---

*This contention is supported by Graves and Oppenheimer (1975).

Providing an occasional hand-out of food, in other words, as an act of benevolence, is neither enlightened compassion nor constructive action.

COUNTRY DOGS

In the field study area in the Nilgiris, the village dogs in Masinigudi consisted of two "societies," one group having homes, the other dogs being homeless scavengers. Both groups, however, were free to roam as they pleased. Those with homes exhibited more territorial behavior (as would be anticipated) and were less gregarious than the homeless dogs who tended to sleep together and had a greater proximity tolerance toward each other. There is an interesting analogy here between the possession of property (or territory) and its effects on the sociability of man and dog alike! Dogs with homes were easily identifiable not only because of a generally (but not invariably) superior physical condition, but also on the basis of their antagonistic reactions toward other dogs of the same sex, especially when close to their homes. How then, do these dogs form hunting packs? Presumably, when off home base (on "neutral" ground) and with the socially facilitating stimulus of the prey, proximity intolerance is reduced. These dogs are then socially flexible enough to form an effective temporary hunting unit. Some of these village and farm pi-dogs supplemented their diet of scraps and garbage by foraging for food in the surrounding jungle. They would scavenge for carrion and quickly arrive on the scene after whistling dogs had made a kill. Circling vultures and vocalizing jungle crows probably gave them the cue as to the kill's location. How the birds arrived so quickly, even during a hunt, remains a mystery. Perhaps the sight of prey and predators in pursuit and the screams of the prey during attack are detected by the attentive birds. This information is then communicated to other birds and, coincidentally, to scavengers such as hyenas, jackals, and pi-dogs.

Pi-dogs have been reported actually hunting with the chennai. They will also hunt deer by themselves. We observed such hunts on three occasions within a single week. In one incident, three pi-dogs (the largest probably weighing no more than 35 pounds) were seen around noon chasing a small herd of twelve deer. A fourth dog was seen lagging behind and tracking, having lost contact with the main group. So far as we could tell, no kill was made; the dogs were heard barking in deep cover, and langurs were sounding their alarm call. The second occasion involved a yearling chital that ran into the camp (deer will reportedly run toward human habitation when at bay, possibly to seek refuge). The prey had a large laceration on its left foreleg and was clearly exhausted. Two pi-dogs were seen close by, obviously following the deer's tracks. A short while later (3:15 P.M.), the chital was heard screaming in the river. It was being held at bay by one of the dogs, who bit at its rump and neck, even crawling onto the deer's back in order to bite at it. Eight vultures quickly gathered and began circling at 3:30 P.M. and several crows were calling nearby. The chital kept the dog off with a face-to-face frontal threat, but fear was its weakness. Whenever the prey turned away, it would be attacked. The deer eventually found cover in dense brush, and the dogs gave up the hunt.

Three such observations within a single week suggest that village pi-dogs have considerable impact on wildlife, probably maiming more deer than they actually kill.* In addition to this problem, pi-dogs compete for food resources with jackal, hyena, wild dog, and other scavengers and predators. Part of wildlife management must, therefore, focus on the pi-dog problem. Since responsible ownership is negligible, population control measures are indicated. The pi-dogs' impact on wildlife is not generally recognized, and many would implicate the wild dogs for any and all depredations on "game" animals such as chital and sambar. It is intriguing that pi-dogs do not kill domestic livestock, at least to our knowledge. Clearly, if they did not discriminate between wild and domesticated animals, their presence in the villages would not be tolerated by surrounding farmers.

We had not fully realized the extent of the pi-dog's role in prey depredation until our last twenty-four hours in the field. Our final dinner consisted of roast curried sambar (a change from the daily curried vegetable stew or sambhur), given to us by our cook who told us it had been shot by a hunter and given to the villagers. Later, we learned that the sambar had in fact been killed by village pi-dogs. The next morning, while driving out of the valley, we found a group of gypsies cooking by the roadside with a dozen or more dogs tied up on the bridge. The dogs were being taken to Mysore for sale to villagers for the sole purpose of hunting. Villagers will allow their dogs to roam free and will encourage them to chase and kill deer: a good hunting dog is a prized possession. At the official level, this practice, like the grazing of low-yield cattle and buffalo, is accepted as an unavoidable fact of life; the villagers need meat. It is indeed unfortunate that they do not eat their cattle instead. Cattle and buffalo, even when producing no milk, are a status symbol, costing nothing since their feed comes from the jungle. This parasitic exploitation of the jungle's resources by man via his cattle, dogs, and inefficient agricultural practices is not unique to this study area, or to India, but is common in all relatively overpopulated and underdeveloped countries. The global picture is indeed gloomy for the future of the world's wildlife.

Because of their frequent close associations with jackals and wild dogs, village pi-dogs (which probably have greater resistance to distemper and other viral diseases than their wild cousins) could severely reduce the populations of these canids during epidemic outbreaks. Pi-dogs will occasionally crossbreed with jackals (as will feral dogs with coyotes in the United States), but there are no known cases of pi-dog/chennai crosses.

---

*Pi-dogs will often make a kill near a village or patti. They are not discouraged from doing this by the villagers, who will drive the dogs off and take the meat for themselves.

Figure 9.2 Three village pariah dogs scavenging chital remains from a whistling dog kill.

The disease factor, together with direct food competition and their obvious depredations on game animals, makes the rural pi-dog a menace to wildlife management and conservation. These same factors are also present in the United States, where free-roaming and feral rural dogs (often running in packs) cause untold damage to domestic livestock and wildlife.

The entire problem is further aggravated by a constant production of pups from the village and surrounding pattis (farms). Marginal human support buffers the dogs from the population controls to which wild canids would normally be subjected. With such an advantage, the unremitting impact of pi-dogs on wildlife and direct competition with wild canids must be considerable.

Oppenheimer and Oppenheimer (1975) report that pup mortalities are around 33 percent, a figure comparable to the mortality rates of wolf cubs and rural feral dog pups in North America. They also found a peak in estrus activity in pariah dogs around September and October, suggesting a single annual heat characteristic of the dingo, whistling dog, Indian wolf, and African basenji.

A straightforward collar and tag licensing program for all owned dogs in villages close to wildlife sanctuaries would be a first step in solving the pi-dog problem (together with rabies vaccinations, rarely done in India, and, ideally, spay-neuter programs as well). Unclaimed, unlicensed dogs would be humanely destroyed if not adopted, and all owned dogs would be restrained or destroyed if they did roam and hunt. According to one official, if it were not for the price of ammunition, the Indian Forest Service in the Nilgiris would shoot any pi-dogs seen within the sanctuary or caught chasing deer.

The use of poisoned bait (such as Follidol) by farmers to kill marauding leopard and wild dog is now being carefully controlled. Such practices more usually cause the death of vultures, jackals, hyenas, and also pi-dogs. Since these dogs also forage in the villages, the pi-dog population would be least affected by such indiscriminate poisoning so the dogs could easily fill the food-niche temporarily vacated by these other wild species, and fill it so effectively, in fact, that the hyena and jackal especially might never regain a viable population. The most stringent measures are, therefore, needed in the use of poison baits and in the population control of pi-dogs both of which can have a synergistic and adverse effect on wildlife and the ecosystem.

Chapter 10

MAN, NATURE, CONSCIOUSNESS, AND CONSERVATION*

There is something wholly tangible but almost ineffable in the nature of wilderness that can deeply affect the human mind. For me, it is an intuitive sensing of infinity and eternity in the diverse patterns of an omnipresent and transcendent intelligence, manifested as co-creative harmony. Balance is maintained within microcosms that are integrated and interdependent, making one indivisible whole: Nature. It is a totality expressing eons of evolution within which each life and death is part of a continuum of transformation in space and time. At every moment, this continuum forms a unitive experiential whole: the Tao of nature.

Feeling at first overwhelmed and insignificant, my Western, Cartesian, scientific mind-set gradually began to give way and to realize that it is part of a unified field of being in which all is significant and precious beyond measure. Within the many subjective, aesthetic, and even spiritual experiences and impressions of nature, one may catch a glimpse of eternity. With the mind attuned to and thinking in nature, we have what Gregory Bateson† referred to as the necessary unity of mind and nature which may well be the natural state of human consciousness. Poet Gary Schneider

*Portions of this chapter are from the author's *One Earth One Mind,* (Fox 1980*a)*.

†*Mind and nature: a necessary unity*. (New York: E.P. Dutton, 1979).

has eloquently linked this mental unity with meditative states of za zen or Samadhi. Such is the natural mind of the hunter, and, I believe, of naturalists, today's nature lovers, and curious ethologists whose minds must, by necessity, be at one with nature, the Tao of man.

Alf Wannenburgh in *The Bushmen** vividly describes the applied science of ethology as practiced by these African hunter-gatherers, in many respects comparable to the tribal hunter-gatherers roaming through our study area in the Nilgiri hills.

> Bushmen acquire an extensive knowledge of animal behavior through constant observation, careful attention to detail and continual discussion among themselves of what they have seen. Their understanding enables them to identify completely with the animal they are hunting, so that they can answer questions such as: "What would I do now if I were this animal?" And their replies are amazingly accurate. Such anthropomorphic projection is rejected by scientists, who say it is not possible to equate human and animal consciousness, but perhaps it comes more easily to men who believe that all the animals were once people like themselves.

Thanks to science, we have progressed materially and have enhanced our understanding and power over Nature. But, in the process, we have objectified Nature and by so doing are losing that intuitive and spiritual connectedness and sense of reverence our ancestors possessed. This sense of reverence, a pantheism or panentheism, if you wish (minus the baggage of animism and superstition), is needed to temper our destructive dominion over Nature.

As Sioux medicine man Black Elk observed, "Nothing can live well except in a manner suited to the way the Power of the World lives and moves to do its work."† The intrinsic worth of a wild dog, banyan tree or a lake is neither perceived nor appreciated when *Homo sapiens* seeks more control and power over the Power of the World, out of greed or sheer necessity.

Perhaps the greatest test of compassion and understanding and the greatest demonstration of dire human need can be found in the "human jungle" of the city, a very different reality from the jungle wilderness. Few can escape from this "jungle", and, when experienced, it can have a profound effect upon your views of humanity and your attitudes toward life and suffering. On the second field trip, we took time to study the homeless, free-roaming pi- or pariah dogs of Indian villages and cities, and saw another side of life. We visited Bombay, Delhi, Srinigar, Madras, and countless small villages before and after my two field studies in the Nilgiris. The smaller the town, the greater the poverty, and the bigger the town, the more the extremes between wealth and poverty were apparent. A heap of

---

*New York: Mayflower books, 1979.

†*Black Elk Speaks* by J. Neihardt (Lincoln Nebraska: University of Nebraska Press, 1979).

skin and bones in the street could be a scabious dog, a dried-out cow, or an emaciated beggar. Like the jungle, the interdependent will to survive was palpable, and survival of the fittest was the rule. But the difference was the degree and extent of overt suffering and sickness, for, in the jungle, sickness and suffering were quickly and mercifully terminated by predator species like the wild dog.

I reflected upon Garret Hardin's essay entitled "The Tragedy of the Commons"* where he showed there are limits to the extent to which resources can be shared. His controversial book, *The Limits of Altruism: An Ecologist's View of Survival* (1977),† showed how acts of benevolence can only serve to prolong life and suffering as resources become even more scarce. Amidst the great temples and shrines of India's rich pantheistic spiritual heritage, the tragedy of reality seemed even more unbearable; and the social, economic, and ecological problems of this great nation, that historically has given so much of its philosophical wisdom and cultural riches to the West, seemed almost insoluble.

Distinctions blurred between the plight of animals and humans: an overloaded and sagging bullock and its undernourished and overworked driver or *tongawalla*; children running behind a cow to collect handfuls of its dung so they might have fuel for their evening meal. The political, economic, and ecological interdependence of animal and human were more tangible here than in the jungle, yet the destiny of both were caught in the same web of fate. In spite of dire human need, and although pariah dogs were shunned and often stoned for fear of rabies,‡ there was an all pervading sense of reverence for life, incongruous as it might seem. For example, a placid cow, recumbant and chewing her cud on a main thoroughfare in Madras, was an island unto herself as the traffic sped around her. The jungle, too, was an island filled with nature's riches, that may soon be lost to the world forever.

Western visitors to India must try to see what the people have, and not focus simply upon what they lack materially. We may then realize that our own materialistic and affluent culture has a poverty of spirit which stands in sharp contrast to the vital and unifying spiritual dignity beneath the overt deprivation and suffering of overpopulated India. Wise, yet controversial, government educational programs to promote responsible popu-

---

**Science* 1968 *162*:1243–48.

†Indiana University Press, Bloomington Indiana.

‡According to a report in *USA Today* (June 9, 1983), India has an estimated fifteen million stray dogs and has the world's highest incidence of rabies (with 700,000 people being bitten annually by rabid dogs). Yet, ironically, of Bombay's five million homeless street dwellers "some of whom compete with dogs for food at garbage cans," city dogcatchers complain of being attacked by slum dwellers and merchants when they round up dogs for destruction. "Recently 300 street dwellers gave a ceremonial funeral to a stray dog that had shared their pavement until killed in a road accident."

lation control, conservation, and a self-sustaining agricultural system give grounds for hope. To motivate for profit—to produce commercial surpluses beyond immediate needs—will only hasten the destruction of the jungle, of all wildlife, wilderness, and potential resources of India in one generation.

These observations and experiences provide a poignant lesson on how we may react to the economic, social, and ecological crises our industrial civilization faces today. They stimulate a critical appraisal of human values and goals. Solving these problems and averting a global nemesis may be beyond the grasp of our generation, but they must be recognized and faced today and not rationalized or denied by government and industry. If there is a solution within our reach, it must be incorporated within our consciousness, which embodies our beliefs, hopes, and values as well as directs and determines our actions and priorities. Without a reverence for all life, an ethic of responsible and compassionate action, there can be little hope for whistling dogs or for the majority of future generations of our own kind. Speaking in support of animal welfare, rights, and conservation in the face of human poverty and suffering may seem hypocritical, but, in reality, the interests of animals and humans are inseparable. Christmas Humphreys in *The Buddhist Way of Life** gives the following relevant insight:

> Life is one, said the Buddha, and the Middle Way to the end of suffering in all its forms is that which leads to the end of the illusion of separation, which enables man to see, as a fact as clear as sunlight, that all mankind, and all other forms in manifestation are one unit, the infinitely variable appearances of an indivisible Whole.

ANIMALS BEFORE PEOPLE?

It is considered heroic altruism for men to help women and children first onto the lifeboats of a sinking ship. But what would Noah have done if he, his family, and the world's menagerie of creatures were on a sinking Ark? Norman Myers in his book, *The Sinking Ark,* shows very clearly that both we and our animal kin are in grave danger of extinction, and we have no lifeboats or safe place to go. Furthermore, our heroic altruism to save our own kin and species, as Garret Hardin explains in his book *The Limits of Altruism* can be misguided. In valuing human life over the "rights" of other living beings and in ignoring sound ecological principles,† we bring a glo-

---

*New York: Schocken Books, 1969.

†These principles include what he terms the "carrying-capacity ethic" of not living or multiplying beyond nature's capacity to sustain our needs as well as the needs of other living things essential for the maintenance of a healthy environment.

bal nemesis upon ourselves. As Lester Brown documents in *The Twenty-Ninth Day* (1978)* we are running out of time.

Albert Schweitzer (1965) once observed, "By ethical conduct towards all creatures, we enter into a spiritual relationship with the universe." René Dubos (1972) added an environmental dimension in *A God Within* (which Aldo Leopold presaged in *A Sand County Almanac*)† when he stated, "A truly ecological view of the world has religious overtones. . . . An ethical attitude to the scientific study of nature readily leads to a theology of the earth." And he wisely cautions:

> A relationship to the earth based only on its use for economic enrichment is bound to result not only in its degradation but also in the devaluation of human life. This is a perversion which, if not corrected, will become a fatal disease of technological societies.

Humanitarians and conservationists, when voicing concern over the fate of animals, wild and tame, are seen by their critics as placing animals before people. Clearly, in their critics' minds, this concern is anthropomorphic and sentimentally misguided altruism, since no rational person would value animal life over human life. Biomedical scientists and spokespersons for various industries, such as agriculture, energy, and forestry, are especially critical of humanitarians and conservationists. The former individuals believe that the exploitation of animals and ecosystems is for the good of humanity and, even though their activities are profitable, their endeavors are intrinsically altruistic and, therefore, justifiable.

A dialectical tension is evident between our need to exploit and destroy life to sustain our own, and the desire to respect, cherish, and preserve all life from needless exploitation and suffering. Some moral philosophers, such as Regan (1982), Singer (1977), and Rollin (1981), have argued that animals have intrinsic value and that, as yet, no empirically demonstrable, morally relevant differences between humans and other animals have been found to justify their wholesale exploitation for often trivial purposes. Might (dominion expressed as domination) does not make Right. (See also Fox 1980b and Morris and Fox 1978.)

But no matter how articulate and airtight these moral arguments may be, most people remain convinced that animals are inferior. To suggest that animals should be given equal and fair consideration is interpreted as placing animals before people. Yet there are no biologically or morally relevant reasons for not giving animals of the same degree of sentience, equal and fair treatment. The misinterpretation and rejection of animal rights philosophy, conservation, and humane ethics is widespread, but understandable.

---

*W. W. Norton, 1978.

†London: Oxford University Press, 1977.

Our cultural values are anthropocentric and our intellectual and technological superiority over other animals gives us reason to believe that we are, in certain ways, superior. To feel that our own children are not special and worth dying for, if need be, is unthinkable, and, as sociobiologists have shown, without such altruism we would surely have become extinct, or at least never have become civilized, long ago.

The value and virtue of moral, animal rights philosophy (akin to a trans-species altruism), the ecological ethics of Rodman (1974) and Stone (1974), and the humane, reverence-for-life ethics of Albert Schweitzer (1965) will never be realized until the "animals before humans" versus "humans before animals" polemic is reconciled. Intellectual argument, an appeal to reason or to higher moral and spiritual values, can be too easily rationalized and denied on the grounds of utility and necessity. Necessity justifies all forms of animal exploitation and environmental manipulation and destruction, regardless of the initial degree of necessity. Moral argument is then relatively futile, especially in cultures that place economic values before ethics or in countries facing overpopulation and starvation. Animal and environmental rights philosophy tends to foster self-righteous moralizing, moral superiority, and a judgmental attitude toward opponents which only increases the existing polarization. Furthermore, using intellectual argument has serious limitations because it simply meets counter-intellectual rationalizations, denial, and belief. No matter how excellently presented, environmental and animal rights philosophy will continue to be resisted by rationalizations, such as the economic justification of all animal exploitation as necessary for human good, and by beliefs, such as humans are superior, have immortal souls, God-given dominion, etc.

Rather than appealing simply to reason, therefore, a nonphilosophical approach may be more fruitful. Reason alone may convince a few of the values and virtues of animals rights philosophy and environmental ethics (and of the long-term economic advantages in conservation and pollution control), but, as Pascal observed, "The heart has its reasons which reason knows nothing about." The human heart has a deeper wisdom, which is not sentimental nor rational, but rather, is intuitive and empathetic. But, as psychotherapists have long recognized, this intrinsic part of our being is cut off by defensive rationalizations and denial. Ideological hypnosis (the Orwellian unconsciousness of Orthodoxy) and appeal to reason only reinforces these mechanisms causing further rationalization and denial. We continue to justify the destruction of the planet and the cruel exploitation of life, and to deny that the "Ark" is indeed sinking. As socioeconomic, environmental, and human health problems increase, the more intensely we seek refuge from anxiety behind such denial and rationalizations.

However, we need not moralize philosophically, but simply open our eyes and hearts to the world around us for one moment and courageously and honestly experience the just-so-ness of reality. And, in the process of awakening our powers of intuition and empathy, we may realize that there is *no* difference in considering animals before people or vice versa, for as the Book of Ecclesiastes (3:20) states, ". . . a man hath no preeminence above a beast: for all is vanity." Indian spiritual teacher Sadhu Vaswani expressed it this way:

> The beast and the bird cannot speak to us in the language we understand: They cannot protect themselves: for centuries they have suffered for our sins against them. The blessed Buddha said: "When wisdom came to me, I resolved to defend the weak, and to all living things I gave compassion of my heart." Yes, with wisdom grows the compassion of feeling Kinship with all life.

If we are superior to animals, then we must have humility. If we must exploit them and nature, then it must be done sparingly and humanely. Our anthropocentric world view must become "ecocentric" (think globally and act responsibly locally), and our altruistic concern must embrace all species, not simply our own. This change in heart and mind is a survival imperative for our species and for our planet alike. W. Jackson Davis, in *The Seventh Year,** concludes that

> Ecology has revealed the intrinsic logic of Karma and the Golden Rule: what we do unto nature, we do unto ourselves as well. We are in a position to understand that we do not *control* nature: we *are* nature, and it is us. There is no separation: all is one, propelled through different physical forms in space-time by nature's omnipresent cycles. The knowledge generated by the Industrial age has paved the way for a historic fusion of the material and spiritual planes.

Animals before people? Yes, in most instances: this attitude is the spiritual, humane, and ecological basis for our survival. People before animals? No, except in those exceptional instances where their exploitation is essential for our survival (or for the balance of ecosystems). Such necessary exploitation, tempered by the wisdom that recognizes the human survival value of considering animals before people, makes us not only humane, but more likely to survive and prosper in love and harmony with all life.

In the ancient Japanese religion of Shintoism, which predates Buddhism, there is the saying, "The awe inspiring Deity manifests Itself in all things; even in a single leaf of a tree or a delicate blade of grass." Lao-Tzu,† the founding philosopher of Taoism, proclaimed, "To nourish one's spirit is to return to nature." He also observed:

> Those who would take over the Earth and manage it
> I see they cannot grasp it;
> for the Earth is a spiritual vessel

---

*New York: W.W. Norton, 1979.

†*Tao tê ching* Ch'u a-kao (transl.) (New York: Samuel Weiser, 1973).

and cannot be forced.
Whoever forces it spoils it
Whoever grasps it loses it.

RETURN TO THE JUNGLE

The primordial and pristine jungle restored our spirits when we left the urban jungle behind. After the rains, macrocosm and microcosm were suffused with a numinous vitality and the wild dog's world seemed even more magical in its beauty and completeness. Poetic images crystallized: a fig tree swayed, its dappled shadows pulsing like a feasted leopard purring in the shade. Along the river, golden leaves of crisp bamboo made royal carpets for deer and peacock. Festive voices and a temple drum in noontime breezes rose and fell; the jungle sighed, a hum of flies. Its forest glades enchant unwary minds—beware the elephant and golden hinds!

The age of the dinosaurs, giant reptiles and insects, and the age of the great mammoths are past. The age of mammals has become the age of Man, and yet, in this valley in the Nilgiris each dawn and dusk resounds with myriad calls of insect, bird, and mammal; the grasses and leaves add to these incessant rhythms and pulsations. The sacred Banyan tree by the river seems to have been listening, growing, and waiting. Perhaps the tree waits for us to recognize that there is no ineffable mystery, no "higher" purpose to life, other than the expression and fulfillment of life itself (and not simply human life at that). For the Hindus of India, Eternity is here with us now—we are, at this moment, in eternity as much as we ever shall be, and our present life is but one in an infinite number of fleeting moments in the eternal life.

In the existential presence or "eternal now" of Nature is the arbitrary "just-so-ness" of reality, ordinary yet ineffably transcendent and divine in its inspiration and conception. Awareness of this reality is experienced as a subjective sense of oneness of mind with Nature, without the intrusion of objectifying thought. Such thought modalities (as preconceptions, expectations, etc.) sever what is perceived from the perceiver, breaking the sense of unity that exists when the perceiver and what is perceived are one. This same state of mind is experienced by many when they perceive and contemplate some object or fellow being whom they cherish, revere, and love. It is that state of being-in-awareness philosopher Martin Buber termed "I–Thou."

Such a state of being is not enchantment or a figment of narcissistic, symbiotic yearning. Rather, it bridges the dialectical tension between subject-object manifolds (as between the subjective "I" and the objective "You") with the realization of the just-so-ness of one's being in Being, as the oneness of I–Thou.

This oneness is of connectedness and kinship (as objectively conceived by the sciences of ecology and evolution), wherein one's individuality is not lost (as it is in symbiotic regression), but enhanced (suprapersonalized, according to Teilhard de Chardin) by the objective "otherness" and subjective just-so-ness of rocks, trees, rivers, deer, wild dogs, and the entire *umwelt* of one's consciousness. From this perspective, Mind and Nature are one, the *anima mundi* and *anima* and *animus* of the human psyche or soul forming a unified field, a holographic ecology of mind.

Alan Wittbecker* provides etymological evidence of this oneness and connectedness with Nature as follows:

> The body is the primary unit; it is embodied in the world, embodied in being. From the oldest language we know, the Indo-European tongue, we took the word for earth *(dhghem)* and turned it into humus and human (dhghem = earth → humanus in Latin → human in English). Yet, the word for man was shaped into man-image, world (Indo-European *wiros* = man → weorold in Old English → world). One word progresses from earth to human, the other from human to earth. We cannot be any closer to the earth and its processes, since the parts are combined in us. We are indissolubly one with nature.

Being is Mind-at-large, immanent in Nature's intelligent organization and in the evolving sentience and sapience of individuated beings, yet also transcendent. This Mind is mirrored like an enantiomorph in the Self that is also immanent in the subjectivity of one's relationships, and yet, is transcendent in the objective consciousness of a separate observing ego or soul.

As Mind, Self, and Ego are one, so is the Self and Nature (the Soul of Nature or *anima mundi)* when the duality of self and other is perceived as a perceptual and cognitive artifact of life's dialectical materialization. This "necessary unity between mind and nature" as the late Gregory Bateson termed it, represents a gestalt or paradigm shift in awareness from objectifying Nature (as something "other" or alien), to fully subjective (empathetic) sentience and objectively sapient consciousness. Martin Buber† termed this state of being in relationship, I–Thou.

Sentience and sapience increase in intensity or vibration in Nature's individuated evolution which, in us, is manifested as empathetic compassion and understanding, and intuited as agapé and Logos, as one enters "the clear cold light of love and life." This numinous view of Mind and Nature, and the qualitatively different kind of attitude and relationship toward Nature that this view engenders is necessary, I believe, to guide and con-

---

*"One Earth, Many Worlds: Prelude to a Holocosmology," Ph.D. dissertation, International College, Los Angeles, 1983.

†*I and Thou* Walter Kauffmann (transl.) (New York: Scribners, 1970).

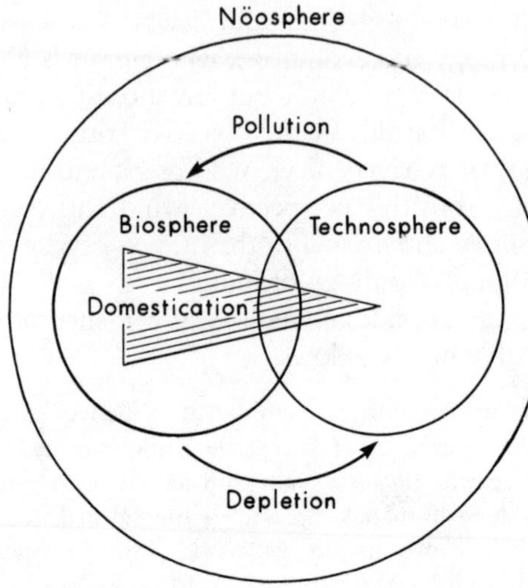

Figure 10.1 The biosphere, the balanced world of nature, is depleted and polluted by the unbalanced technosphere of man and further modified by the invasion of domestic plants, animals, and their diseases. The technosphere must become self-contained and balanced in terms of energy requirements, otherwise the biosphere will be irreversibly destroyed. Awareness on a global scale, where shared values and concerns unite humanity, creates the nöosphere. Our technology and the crises of the biosphere may therefore, in the long run, have a uniting effect on humanity.

strain how we treat all of creation. A purely objective ("I–It"), materialistic, and dominionistic mind-set is self-limiting and, ultimately, self-destructive.

With these thoughts in mind, I hope that this study of the ecology and behavior of the whistling hunters has provided sufficient scientific evidence and inspiration to validate the spiritual reasons for the conservation of Nature and all species of life. In its fullest sense, according to the Christian (and panentheistic) view of the Very Reverend James Morton (Dean of the Cathedral of St. John the Divine, New York City), "Ecology is the science of the Body of Christ through which we of the earth community learn our sacred connectedness." Ecology thus leads naturally to the revelation of the whole, and thus to that which is holy.

This is the essence of the "deep ecology" and animal rights movements today. Tomorrow, such perceptions and ethics will hopefully be recognized as a shared, common morality of enlightened self-interest. All may then enjoy the communion of reverence for all life in a world restored to its natural beauty and physical and spiritual health. This is a consummation devoutly to be wished for the good of all creatures and our children's children alike.

In the preservation of Nature is the preservation of the "otherness" of ourselves, for we are an inseparable part of the biosphere—a unified field of being and becoming. From this perspective we may find the inspiration,

Figure 10.2 A whistling dog surveys its jungle world. Neither may exist in twenty years.

wisdom and humility so essential for our harmonious participation in the creative process of planetary and personal evolution.

CONCLUSIONS

Although wild dogs are hunters and, therefore, killers, like wolves, they show a high degree of sociability and little aggression within their ranks. This finding is contrary to popular opinion, since it is generally held that a hunter must be aggressive. The same misconception has been applied to man whose ancestry as a hunter is thought to be the source both of an innate aggressive urge and of our present ills and destructive mode of being. This contention is not supported by what is known of the biological and social heritage of human and nonhuman hunters. We should therefore look for a more recent source for man's propensity to do violence to others: this source is cultural. There is nothing wrong with human nature. What is wrong is our nurture: our enculturated values, needs, and perceptions.

Alternative and appropriate values, resources, technologies, beliefs, and life styles are urgently needed—not to stop progress, but to bring about a steady state which *is* progress. We must now assume the awesome and godlike responsibility of the earth's destiny, a role into which we were inevitably cast when our species first began domesticating plants and animals 10,000 years ago. It is imperative today that our technology—the technosphere—be modeled like the biosphere, a self-contained, balanced system that neither pollutes nor overexploits and depletes.

I went in search of Wildness and whistling jungle dogs. Although, at its conception, the project was a purely scientific search for knowledge, it became an inner, personal search for meaning. In Wildness I found more than whistling jungle dogs: my purpose in studying these animals led me to an awareness of a greater whole, of which man has become an alienated part. I discovered a lost part of myself. This discovery leads me to propose that within the nature of man is the seed of awareness which can reunite us with the natural world, a necessary unity since the future of both are inextricably bound and interdependent ecologically, economically, and spiritually. Having reverence for all life can enrich our own lives and broaden the scope of our moral concern through compassionate and ethically responsible actions. Such actions will benefit animals and our fellow beings alike, directly or indirectly as the case may be. Ideally, we should think and act in no other way, and, hopefully, one day we will. Such is, perhaps, our greatest challenge while abiding on Earth. Meeting this challenge is as much a survival imperative as it is an evolutionary step forward in our own spiritual development.

"Everything is interdependent. Man, Animal, and Environment, whatever the economical or political context. Everything is related. Whatever happens now to animals will eventually happen to man. The conservation of our inheritance deserves the same natural care as our economical development."

—Indira Gandhi

REFERENCES

Brown, L. R. 1978. *The Twenty ninth day*. New York: W.W. Norton.

Burton, R. W. 1941. The Indian wild dog. *J. Bombay Nat. Hist. Soc.* 41:691–715.

Clutton-Brock, J.; Corbett, G. B.; and Mills, M. 1976. A review of the family Canidae with a classification by numerical methods. *Bull. Br. Mus. (Nat. Hist.) Zool.* 29:119–199.

Cohen, J. 1977. A review of the biology of the dhole or Asiatic wild dog *(Cuon alpinus,* Pallas). *Anim. Reg. Stud.* 1:141–158.

Connell, W. 1944. Wild dogs attacking a tiger. *J. Bombay Nat. Hist. Soc.* 44: 468–470.

Crisler, L. 1958. *Arctic wild.* New York: Harper and Row.

Davidar, E. R. C. 1965. Wild dogs *(Cuon alpinus)* and village dogs. *J. Bombay Nat. Hist. Soc.* 62:146–148.

———. 1974. Observations at the dens of the Dhole or Indian wild dog. *J. Bombay Nat. Hist. Soc.* 71:373–374.

———. 1975. Ecology and behavior of the Dhole or Indian wild dog *(Cuon alpinus*). pp. 109–119. In *The Wild Canids*, ed. M. W. Fox. New York: Van Nostrand, Reinhold and Co.

Dubos, R. 1972. *A God Within*. New York: Scribners.

Fox, M. W. 1971. *The behavior of wolves, dogs and related canids*. New York: Harper and Row.

———. 1978. *The dog: its domestication and behavior.* New York: Garland Press.

———. 1980*a*. *One earth, one mind.* New York: Coward McCann.

———. 1980*b*. *Returning to Eden.* New York: Viking.

Fox, M. W., ed. 1975. *The wild canids.* New York: Van Nostrand, Reinhold and Co.

Fox, M. W.; Beck, A. M.; and Blackman, E. 1975. Behavior and ecology of a small group of urban dogs. *Applied Anim. Ethol.* 1:119–137.

Fox, M. W.; Lockwood, R.; and Shideler, R. 1974. Introduction studies in captive wolf packs. *Z. Tierpsychol.* 35:39–48.

Gee, E. P. 1964. *The wildlife of India.* London: Collins.

Graves, I. L., and Oppenheimer, J. R. 1975. Human viruses in animals in West Bengal. *Human Ecol.* 31:105–130.

Hardin, G. 1977. *Limits of altruism: an ecologist's view of survival.* Bloomington: Indiana University Press.

Johnsingh, A. J. T. 1979. Ecology and behavior of the wild dog *(Cuon alpinus)*. Ph.D. dissertation, Sivakusi College, Tamil Nadu, S. India.

———. 1982. Reproductive and social behavior of the dhole, *Cuon alpinus. J. Zool. (Lond.)* 81:443–462.

Keller, R. 1973. Einige beobacktungen zum verhalten des dekkanrothundes *(Cuon alpinus dukhensis,* Sykes). *Wert. Natur. Fesch.* (Zurich) 118:129–136.

Khajuria, H. 1963. The wild dog *(Cuon alpinus)* and the tiger *(Panthera tigris). J. Bombay Nat. Hist. Soc.* 60:448–449.

Kleiman, D. G. 1967. Some aspects of social behavior in the Canidae. *Am. Zool.* 7:365–372.

Krishnan, M. 1972. An ecological survey of the large mammals of peninsular India. *J. Bombay Nat. Hist. Soc.* 69:26–54.

Kruuk, H. 1972. *The spotted hyena.* Chicago: University of Chicago Press.

Lee, R. B., and DeVore, I., eds. 1968. *Man the hunter.* Chicago: Aldine-Atherton.

Mech, L. D. 1970. *Ecology and society behavior of the wolf: an endangered species.* New York: Natural History Press, Doubleday and Co.

Morris, R. K., and Fox, M. W. 1978. *On the fifth day: animal rights and human ethics.* Washington, D.C.: Acropolis.

Myers, N. 1979. *The sinking ark.* New York: Pergamon.

Novikov, G. A. 1956. *Carnivorous mammals of the fauna of the U.S.S.R.* Israeli Porgram for Scientific Translation, Jerusalem.

Oppenheimer, E. C., and Oppenheimer, J. R. 1975. Certain features in the pariah dog in West Bengal. *Applied Anim. Ethol.* 2:81–92.

Prater, S. H. 1965. *The book of Indian animals.* Bombay: Bombay Natural History Press.

Pythian-Adams, E. G. 1939. The Nilgiri Game Association 1879–1939. *J. Bombay Nat. Hist. Soc.* 31:384–396.

———. 1949. Jungle memories. Part V. Wild dogs and wolves. *J. Bombay Nat. Hist. Soc.* 48:180–240.

Regan, T. 1982. *All that dwell therein.* Berkeley: University of California Press.

Rodman, J. 1974. The dolphin papers. *North American Review.* 12:6–33 (Spring)

Rollin, B. 1981. *Animal rights and human morality.* Buffalo: Prometheus.

Saratchandra, H. C., and Gadgil, M. 1975. A year in Bandipur. *J. Bombay Nat. Hist. Soc.* 72:625–647.

Schaller, G. B. 1967. *The deer and the tiger.* Chicago: University of Chicago Press.

Schaller, G. B. 1972. *The Serengeti Lion.* Chicago: University of Chicago Press.

Schweitzer, A. 1965. *The teaching of reverence for life.* New York: Holt, Rinehart and Winston.

Singer, P. 1977. *Animal liberation.* New York: Random House.

Sosnovskii, I. P. 1967. Breeding of the red dog or dhole *Cuon alpinus* at Moscow Zoo. *Int. Zoo Yearb.* 7:120–122.

Spillett, J. J. 1966. A report on wildlife surveys in South and West India. *J. Bombay Nat. Hist. Soc.* 65:633–663.

Stone, C. 1974. *Should trees have standing?* Los Altos, CA: Wm. Kaufmann.

Teilhard de Chardin, P. 1970. *Man's place in nature.* New York: Harper and Row.

Thenius, E. 1954. On the origin of the dhole. *Osterr. Zool. Z.* 5:377–388.

Van Lawick, H., and Van Lawick-Goodall, J. 1970. *The innocent killers.* New York: Houghton Mifflin Co.

Waller, R. H. 1972. Observations on the wildlife sanctuaries of India. *J. Bombay Nat. Hist. Soc.* 37:949–950.

Wynne-Edwards, V. 1962. *Animal dispersion in relation to social behaviour.* London: Oliver & Boyd.

INDEX